Mechanics and Adaptronics

Ziel der Buchreihe ist die Publikation neuer Ansätze in der Mechanik und Adaptronik. Diese Ansätze umfassen zum einen experimentelle Themen als auch solche, die numerischen Charakters sind und erstrecken sich über diverse Größenskalen. Experimentelle, zukunftsweisen Themen beinhalten Untersuchungen auf Mikro- bis hin zur Makroebene und berücksichtigen unterschiedlichste Felder, von der Mechanik über Temperatur bis hin zu unterschiedlichen chemischen Feldern. Auch auf Seiten der numerischen Simulation werden auf Grundlage der experimentellen Untersuchungen neue multiskalige Mehrfeldmodelle entwickelt, kalibriert und validiert. Eine Besonderheit stellt hier die Thematik der Adaptronik dar, die als Ziel die Integration neuer Funktionen in tragende Bauteile durch die Kombination konventioneller Konstruktionswerkstoffe mit aktiven Werkstoffsystemen verfolgt. So werden Funktionswerkstoffe betrachtet, die Energiewandler sind und die die klassischen lasttragenden und formgebenden Funktionen um sensorische und aktorische Funktionen erweitern. Aktorische und sensorische Eigenschaften ermöglichen in Kombination mit adaptiven, schnellen Reglern eine optimale Selbstanpassung an die jeweilige Einsatzumgebung. Durch die Zusammenführung von numerischem und experimentellem Know-how auf mechanische als auch adaptronischer Seite ist es möglich, Erkenntnisse aus der Natur im Sinne des bionischen Gedankens in die Technik zu übertragen und so neue wissenschaftliche Richtungen aufzuschlagen.

Martin Liebisch

Design Load Allowables for Composite Plates Exposed to Thermomechanical Loads

Springer

Martin Liebisch
Braunschweig, Niedersachsen, Germany

ISSN 2731-6211 ISSN 2731-622X (electronic)
Mechanics and Adaptronics
ISBN 978-3-031-71351-4 ISBN 978-3-031-71352-1 (eBook)
https://doi.org/10.1007/978-3-031-71352-1

This Springer imprint is published by the registered company Springer Nature Switzerland AG
The registered company address is: Gewerbestrasse 11, 6330 Cham, Switzerland

Design load allowables for composite plates exposed to thermomechanical load conditions

Von der Fakultät für Maschinenbau

der Technischen Universität Carolo-Wilhelmina zu Braunschweig

zur Erlangung der Würde

eines Doktor-Ingenieurs (Dr.-Ing.)

genehmigte Dissertation

von: Martin Liebisch

aus (Geburtsort): Jena

eingereicht am: 11.05.2023

mündliche Prüfung am: 08.11.2023

Gutachter:

Prof. Dr.-Ing. Michael Sinapius

Prof. Dr.-Ing. habil. Raimund Rolfes

2024

Contents

About the Author

Martin Liebisch is working at the German Aerospace center (DLR) since 2014. His work was initially focusing the analysis of composite curing analysis. Later his scope changed to the topic of thermomechanical analysis and the behavior of composites under fire exposure. Futher work focuses on numerical and experimental investigation of hydrogen storage vessels.

Symbols

Scalars

α	Thermal expansion Coefficient
β	Chemical shrinkage Coefficient
Δ^P	Potential
ε	Strain
η	Effective strength ratio
f_W	Weakening factor within Puck's failure theory
γ	Shear strain
λ	Thermal conductivity
μ	Mean value
ν	Poisson ratio
ξ	Degree of cure
Φ	Probability density function
ϕ	Ply Angle
ψ	Analysis uncertainty
ϱ	Density
σ	Stress
τ	Shear stress
θ_{FP}	Fracture plane inclination
A	Area
b	Thermal shape parameter
c_p	Heat capacity
cv	Variation coefficient
E	Young's modulus
e	Error
f^i	Failure index
f^{res}	Reserve factor
F	Force
F_{ax}	Axial Force
G	Shear modulus

H	Property degradation magnitude
I	Moment of inertia
K	Structural stiffness
m	Model parameter
P	Arbitrary property
q	Quantile
Q	Heat flux
R	Strength
r^{lin}	Linearity ratio
s	Standard deviation
t	Time
t	Thickness
t^{ply}	Ply Thickness
T	Temperature
T_g	Glass Transition Temperature
T^{max}	Maximum Temperature
T_m	Melting temperature
T^{uni}	Uniform Temperature
T^{alw}	Allowable Temperature
T^{avg}	Average temperature
T^{cure}	Cure temperature
$T^{\sigma 0}$	Stress-free temperature
u_{ax}	Axial shortening
var	Variance
V_f	Fibre volume ratio

Vectors and Matrices

$\underline{\varepsilon}$	Strain vector
$\underline{\kappa}$	Curvature vector
$\underline{\sigma}$	Stress vector
$\underline{\sigma}^F$	Failure stress vector
$\underline{\sigma}^R$	Residual stress vector
$\underline{\underline{A}}$	Membrane stiffness matrix
$\underline{\underline{B}}$	Coupling stiffness matrix
$\underline{\underline{C}}$	Stiffness matrix
$\underline{\underline{D}}$	Bending stiffness matrix
$\underline{\underline{Q}}$	Reduced stiffness matrix
$\underline{\underline{T}}$	Transformation matrix
$\underline{M}$	Vector of bending moments
$\underline{N}$	Vector of normal forces

Indices and Superscripts

$(\cdot)^{alw}$	Allowable, value used for structural design
$(\cdot)^{bu}$	Buckling load
$(\cdot)_c$	Compression loading
$(\cdot)^{cure}$	Composite curing
$(\cdot)^{eff}$	Effective value
$(\cdot)^{el}$	Value related to mechanical (elastic) problems
$(\cdot)_f$	Fiber
$\hat{(\cdot)}$	Model quantity
$(\cdot)^{max}$	Maximum value
$(\cdot)^{RT}$	Room temperature
$(\cdot)_t$	Tension loading
$(\cdot)^{th}$	Value related to thermal problems
$(\cdot)^{tot}$	Total
$(\cdot)^{tp}$	Training point
$(\cdot)^{ult}$	Ultimate value
$(\cdot)^{uni}$	Uniform

Acronyms

BC	Boundary Condition
CE	Cyanate Ester
CFRP	Carbon Fibre Reinforced Plastic
CLT	Classic Lamination Theory
CP	Cross-Ply laminate
CS	Certification Specifications
CTE	Coefficient of Thermal Expansion
DiA	Distributed temperature Analysis procedure
DMA	Dynamic Mechanical Analysis
EASA	European Aviation Safety Agency
FEA	Finite Element Analysis
FF	Fiber Failure
FPF	First Ply Failure
FRP	Fibre Reinforced Plastic
FSDT	First-order Shear Deformation Theory
IFF	Inter-Fiber Failure
ILSS	Interlaminar Shear Strength
IPS	In-Plane Shear
KDF	Knock-Down Factor
LB	Local Buckling
LL	Limit Load
LOO	Leave-One-Out
MAP	Modified Angle-Ply laminate
MaxAE	Maximum Absolute Error
MaxRE	Maximum Relative Error
MD	Multidirectional
MS	Margin of Safety
MSC	Maximum Stress Criterion
PDF	Probability Density Function
PES	Polyetersulfone
RE	Relative Error

RMSE	Root Mean Square Error
RMSRE	Root Mean Square Relative Error
RT	Room Temperature
SF	Safety Factor
tCE	Toughened Cyanate Ester Matrix
tCFRP	Toughened Cyanate Ester Carbon Fiber Composite Material
UD	Unidirectional
UL	Ultimate Load
UnA	Uniform temperature Analysis procedure
WWFE	World-Wide Failure Exercise

Chapter 1
Introduction

Lightweight design is a central aspect of many industries. It is highly important for aerospace structures where structural weight minimization can lead to improved economic performance and a better ecological footprint, as for instance by reduced fuel consumption. In parallel to the ambition of weight minimization, the request of improved safety is ever-present. Thermomechanical load conditions are demanding and often require a certain design trade-off regarding weight to ensure safety and thereby pose a significant challenge to designers. These conditions introduce additional complexities to the structural mechanical behavior, as temperature changes can alter the effective properties of materials and lead to thermal expansion causing additional deformation and stress development. These dependencies are critical as they make the already complex structural design process even more challenging. As a result, it is essential to reduce this complexity through reasonable simplifications that ensure a safe design while still addressing the added thermal load. This can be achieved by applying conservative design practices and simplifying the analysis, while taking into account the thermal effects on the structure.

Thermomechanical load conditions are a common occurrence in aerospace applications. However, the magnitude and distribution of these loads can vary significantly depending on the specific application and structure being considered. For general aircraft, the maximum temperatures can exceed 90.0 °C due to the extremes of ambient air conditions and solar heating [1, 2]. Although a uniform temperature assumption is often used, the temperature gradient due to the difference between internal state and external surface temperature and the presence of structural members cannot be ignored [3]. Aircraft systems, such as electrical units, batteries, and auxiliary power units, can generate additional thermal loads that act on top of the aforementioned loads. For instance, Sanchez et al. [4–6] investigate the criticality of additional heat load generated within the engineering bay. The resulting temperature distribution and magnitude vary clearly over the location of the aircraft section. Moreover, future

M. Liebisch, *Design Load Allowables for Composite Plates Exposed to Thermomechanical Loads*, Mechanics and Adaptronics,
https://doi.org/10.1007/978-3-031-71352-1_1

aircraft structures operating on electrical engines or fuel cells [7, 8] may have to tackle similar or higher thermal loads, which need to be considered in the design and analysis process. Additionally, the auxiliary power unit and engine exhaust streams also generate significant heat loads [9–12], which can result in very local and intense thermal load distributions, such as those found in helicopters [12]. Certain aerospace applications and military concepts have higher thermomechanical requirements. For instance, the civil aircraft Concorde, which operated at supersonic speeds, reached a fuselage temperature of 120.0 °C [13] which thereby defined the design limit temperature. The SR-71 military aircraft faced significant challenges related to aerothermoelasticity and thermal engine heat [14, 15]. Garrick [16] provides an overview of other aerospace vehicles that exhibit similar effects. Re-entry spacecraft also experience fluid-structure interactions, which can result in non-uniform temperature distributions and shock waves [17, 18]. Experimental studies on these effects are conducted by Daub et al. [19, 20] and Gülhan [21]. The X15 research airplane program conducted fundamental designs for such aerothermal effects and their structural interaction [22–24]. In conventional aircraft design, these thermal loads are partially considered through material selection or additional solutions such as heat barriers to ensure allowable material service temperature while maintaining conservativeness. However, these solutions can add weight and cost to the design, which could be reduced by using more effective analysis procedures.

Lightweight design is a crucial aspect, particularly in the aerospace industry. In recent decades, Fibre Reinforced Plastic (FRP) composites have gained significant attention due to their high specific properties. The orthotropic nature of these materials, which is a result of the number of laminate layers and their fiber orientations, allows for tailoring the material properties to meet specific requirements. However, FRP composites present specific challenges when exposed to high temperatures and temperature variations. The polymer matrix systems used in FRP composites are inherently temperature-dependent and commonly degrade at high temperatures. As a result, design guidelines, procedures and analysis simplifications must account for these effects to prevent catastrophic failure. While these considerations ensure the safety, reliability, and conservativeness of the design, they can also lead to over-conservatism, resulting in additional weight and reduced design efficiency. Therefore, it is essential to challenge the current analysis procedures for thermomechanically loaded composite structures.

To fully realize the potential of lightweight design, it is essential to identify and exploit existing but unused load-carrying capabilities. This thesis focuses on composite structures subjected to thermomechanical loads and proposes a novel analysis procedure that considers a higher level of detail to model the influence of temperature. By increasing the understanding of the structural behavior, this approach has the potential to identify and utilize existing structural reserves that are currently unused. Moreover, it can also increase confidence in the analysis, which could potentially lead to adapt margins of safety in design.

References

1. D. Petersen, R. Rolfes, R. Zimmermann, Thermo-mechanical design aspects for primary composite structures of large transport aircraft. Aerosp. Sci. Technol. **5**, 135–146 (2001)
2. W.A. Stauffer, A.M. James. Development of advanced composite structures. NASA. Langley Res. Center CTOL Transport Technol. (1978). https://ntrs.nasa.gov/citations/19780019119
3. E.A. Thornton, Thermal buckling of plates and shells. Appl. Mech. Rev. **46**(10), 485–506 (1993). https://doi.org/10.1115/1.3120310
4. F. Sanchez et al., A Thermal Risk Assessment Approach for the Conceptual Design of Aircraft System Architectures: AIAA 2019. AIAA Aviation 2019 Forum (2019). https://doi.org/10.2514/6.2019-3365
5. F. Sanchez, S. Liscouët-Hanke, Thermal risk prediction methodology for conceptual de-sign of aircraft equipment bays. Aerosp. Sci. Technol. **104**, 105946 (2020). ISSN: 12709638. https://doi.org/10.1016/j.ast.2020.105946
6. F. Sanchez, A.M. Huzaifa, S. Liscouët-Hanke, Ventilation considerations for an enhanced thermal risk prediction in aircraft conceptual design. Aerosp. Sci. Technol. **108**, 106401 (2021). ISSN: 12709638. https://doi.org/10.1016/j.ast.2020.106401
7. X. Huang, Z. Zhang, J. Jiang, Fuel cell technology for distributed generation: an overview, in *2006 IEEE International Symposium on Industrial Electronics*. (IEEE, 2006), pp. 1613–1618. ISBN:1-4244-0497-5. https://doi.org/10.1109/ISIE.2006.295713
8. S. Mekhilef, R. Saidur, A. Safari Comparative study of different fuel cell technologies. Renew. Sustain. Energy Rev. **16**(1), pp. 981–989 (2012). ISSN: 13640321. https://doi.org/10.1016/j.rser.2011.09.020
9. M. Liebisch, T. Wille, Sucohs Project. ECCOMAS Thematic Workshop: Computational Multi-Physics, Multi-Disciplinary and Multi-Data - CM3 (2021)
10. M. Liebisch, T. Wille, Exploitation of high performance structures at elevated temperature, in *Aerospace Europe Conference 2020* (2020)
11. P.M. Cali, Numerical analysis of advanced fighter auxiliary power unit exhaust impingement. J. Aircraft **37**(1), 181–183 (2000). ISSN: 0021-8669. https://doi.org/10.2514/2.2578.
12. P. Cheng-xiong, Z. Jing-zhou, S. Yong, Effects of exhaust temperature on helicopter in-frared signature. Appl. Thermal Eng. **51**(1–2), 529–538 (2013). ISSN: 13594311. https://doi.org/10.1016/j.applthermaleng.2012.09.016.
13. N.F. Harpur, Concorde structural development. J. Aircraft **5**(2), 176–183 (1968). ISSN: 0021-8669. https://doi.org/10.2514/3.43926
14. S. Corda et al., The SR-71 Test Bed Aircraft: A Facility for High-Speed Flight Research. NASA Dryden Flight Research Center, Edwards, California, June (2000). https://ntrs.nasa.gov/citations/20000064011
15. M. Lladoó Bordas, Study of the Lockheed SR-71 structure and reproduction of a 1:20 scale Model. Bachelor thesis. Barcelona: Polytechnical University of Catalonia (2020). https://upcommons.upc.edu/handle/2117/331262
16. I.E. Garrick, A survey of aerothermoelasticity. Aerosp. Eng. 140–147 (1963)
17. E.A. Thornton, P. Dechaumphai, Coupled flow, thermal, and structural analysis of aero-dynamically heated panels. J. Aircraft **25**(110, 1052–1059 (1988). ISSN: 0021-8669. https://doi.org/10.2514/3.45702
18. E.A. Thornton et al., Thermoviscoplastic analysis of hypersonic structures subjected to severe aerodynamic heating. J. Aircraft **27**(9), 826–835 (1990). ISSN: 0021-8669. https://doi.org/10.2514/3.45943.
19. D. Daub, B. Esser, A. G ülhan, Experiments on high-temperature hypersonic fluid-structure interaction with plastic deformation. AIAA J. **58**(4), 423–1431 (2020). ISSN: 0001-1452. https://doi.org/10.2514/1.J059150
20. D. Daub, S. Willems, A. G ülhan, Experiments on aerothermoelastic fluid-structure interaction in hypersonic flow. J. Sound Vib. 116714 (2022). ISSN: 0022460X. https://doi.org/10.1016/j.jsv.2021.116714

21. A. G ülhan, S. Willems, D. Neeb, Shock interaction induced heat flux augmentation in hypersonic flows. Exp. Fluids **62**(12) (2021). ISSN: 0723-4864. https://doi.org/10.1007/00348-021-03336-y
22. R. Reed, J. Watts, Skin and Structural Temperatures measured on the X-15 Airplane during a FLight to a Mach Number of 3.3. (1961). https://ntrs.nasa.gov/api/citations/19630004036/downloads/19630004036.pdf
23. W H. Stillwell, X-15 Research Results With a Selected Bibliography. Washington, D.C. (1965). https://apps.dtic.mil/sti/citations/tr/ADB181235
24. J.D. Watts, Flight Experience With Shock Impingement and Interference Heating on the X-15-2 Research Airplane. Washington, D.C. (1968)

Chapter 2
Problem Description and Focus of Investigation

This chapter provides an overview to the framework of this thesis. Section 2.1 presents the current state of the art procedure used to size composite structures with respect to thermomechanical load conditions. The objective of this work is then presented in Sect. 2.2, where the existing approach is challenged and an alternative procedure is proposed. The objective and key points of these sections are revisited at the end of this work. The analysis procedure developed throughout this thesis is outlined in Sect. 2.3, providing an overview of the different steps involved in the methodology and presenting an outline to the chapters and sections of the thesis.

2.1 Sizing of Structures Exposed to Thermomechanical Loads

One of the final steps in an aircraft development program is the detailed sizing of structures. It involves optimizing the structural setup to achieve predefined targets, such as weight minimization. The sizing process includes constraints and requirements that are developed throughout the aircraft development program, starting from the preliminary design phase where the aircraft configuration is fixed [1, 2]. As the design progresses towards the design freeze, the level of detail increases steadily, including the determination of allowables and definition of design loads. The detailed design considers further aspects, such as manufacturability and inspectability, and conducts the task of optimization, resulting in the final sizing. Throughout these steps, thermal and mechanical load conditions, as well as relevant design environment and boundary conditions, are matured and analyzed. The level of detail addressed by analysis methods increases as the design process progresses, starting from simple predictions and leading to final detailed sizing analyses. Thereby, the analysis procedures take

M. Liebisch, *Design Load Allowables for Composite Plates Exposed to Thermomechanical Loads*, Mechanics and Adaptronics,
https://doi.org/10.1007/978-3-031-71352-1_2

into account simplifications and apply different types of conservatism to ensure for a robust and reliable design. The final step is certification, which is done in accordance with authorities such as the European Aviation Safety Agency (EASA) and their Certification Specifications (CS).[1] Other documents, such as the AMC-20 [7] provide interpretation of these regulations and provide more detail and space for development to the designer.

Designing composite structures that are exposed to thermomechanical loads presents a complex challenge. High temperatures can significantly reduce material properties and the anisotropic behavior of composites can complicate their response. Moreover, thermal strains induced by temperature changes and gradients can interact with mechanical loads, leading to complex stress and deformation patterns that need to be properly accounted for in the design process. To address this challenge, an analysis procedure that incorporates appropriate conservatisms is crucial to ensure a robust and reliable design. This section provides an overview of the current analysis process and conservatisms commonly applied for sizing composite structures subjected to thermomechanical conditions. This sets the context for the research presented in this thesis. The objective of this work is to advance the state of the art in this field and propose a novel approach that may lead to improved accuracy of structural analysis and enables a more efficient design.

Figure 2.1 presents a schematic overview of the common analysis procedure used during the sizing process of a composite structure. The structural analysis, typically carried out using Finite Element Analysis (FEA) , lies at the center of the procedure. It employs specified material properties and applies defined load conditions to predict the resulting structural behavior. An assessment of the resulting structural behavior leads to the evaluation of the design and its potential, for instance with regard to weight or load-bearing capacity. To ensure design robustness, each of the blocks in the figure applies specific simplifications and conservatisms.

The *design loads* are derived from a comprehensive load analysis that considers multiple operational conditions and they typically represent the *worst-case* scenarios that occur over the service life of an aircraft. The maximum mechanical loads are defined as the Limit Load (LL) [4]. To account for unforeseen load magnitudes and related uncertainties, a Safety Factor (SF) of 1.5 is applied, resulting in the Ultimate Load (UL).[2] This SF is applied for aircraft structures and is a historically grown, experienced value that is in consensus with aviation authorities [8]. Similarly, a design load exists for temperature, which is defined as the maximum temperature value T^{max} that can possibly occur over the service life of an aircraft.[3] It is important to note that this design load only refers to the maximum (peak) temperature and

[1] The CS-23 [3] and the CS-25 [4] account for normal aeroplanes and for large aeroplanes, respectively. The CS-27 [5] and the CS-29 [6] address small and large rotorcraft, respectively.

[2] As specified within CS 25.303 [4].

[3] Citation of AMC-20-29 paragraph 6d [7]: „*The peak temperature for composite structures installed in close proximity to aircraft systems that generate thermal energy needs to be identified for worst-case normal operation and system failure cases.*

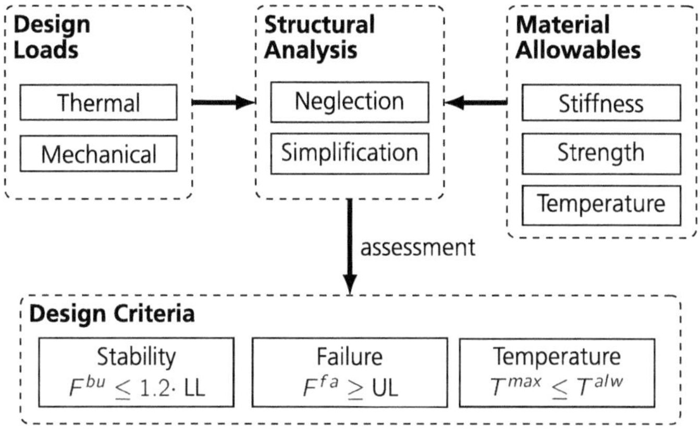

Fig. 2.1 Schematic procedure used for sizing of structures at static thermomechanical load conditions

does not account for the spatial dimension of the heat impingement and resulting temperature distribution. This situation is the primary area of concentration in the current work.

Material allowables are essential for structural analysis and define the magnitude of material properties used in the structural analysis. They are typically determined through experimental programs and consider the statistical uncertainty associated with the material property, as represented by the property's Probability Density Function (PDF). An example PDF for an exemplary material property P at a specified temperature T is shown in Fig. 2.2 (left). The allowable values are defined at the most severe domain of the PDF and are classified into *A-values* and *B-values*, corresponding to the 99% and 90% quantiles, respectively. *A-values* are typically used for primary structures with a single load path, while *B-values* are applied to less severe components with multiple load paths [2, 4, 9]. The allowable temperature T^{alw} of a material is also a conservative limitation and is usually defined by the material's Glass Transition Temperature (T_g). To account for any uncertainties or variations in the temperature conditions, a related Margin of Safety (MS), i.e., a specified temperature margin, is applied to determine the allowable material temperature, as illustrated in Fig. 2.2. For temperature, such values are usually defined empirically, taking into consideration factors such as thermal aging or plasticity onset and typically fall within a range of 30 K to 50 K. Hence, the resulting allowable temperature is the maximum temperature that a structure faces throughout his life and the overall design has to ensure such behavior with respect to worst case conditions. Such MS are implemented to account for unforeseen conditions or imperfections, which may not be accounted for in the structural analysis due to simplifications. It is worth mentioning that there is strong interaction between moisture and temperature, as

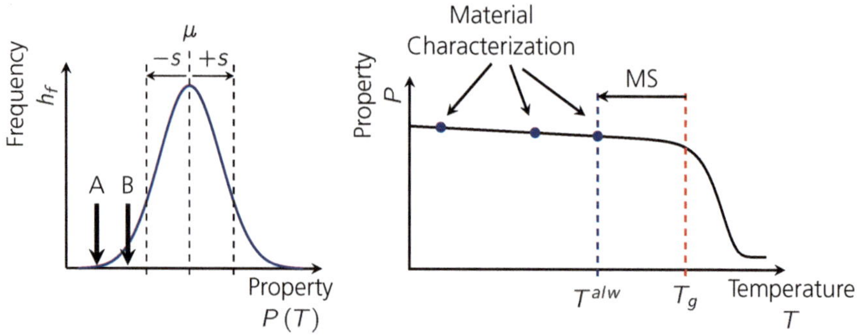

Fig. 2.2 The current sizing philosophy is based on material properties at deterministic temperatures. Left chart: Definition of allowable properties at a specific temperature. Right: Definition of allowable temperature for an exemplaric material. Material characterization is performed at multiple conditions below the maximum allowable temperature

indicated in studies such as [10–12], which should not be neglected. Such environmental conditions and the potential detrimental effects acting on the material[4] are typically considered by using reduced properties [13]. However, this is not further referred to within the scope of this work.

Design Criteria are structural requirements that define the limit state between loads and structural performance. They are implemented to ensure the reliability of structural designs. While the structural behavior resulting from analysis contains all inputs and conservatisms discussed above, the evaluation of the design criteria is the final step that determines whether a structural design can be accepted or not. Figure 2.1 shows the three design criteria that are of interest for the scope of this work. The first crition is given by the temperature requirement, which is already discussed above, and defines that the maximum service temperature must be below the allowable material temperature. This criterion is usually investigated within an individual thermal analysis. It is not explicitly scrutinised or discussed in the context of this work, but it is an important parameter for design of thermally loaded structures. The second criterion is stability bifurcation, also known as buckling, which is a critical structural mechanical phenomenon that describes abrupt changes in the deformation state. Over the past few decades, advances in prediction methods have led to an improved understanding of structural post-buckling behavior and increased confidence in stability analysis [14]. This has resulted in more lightweight designs by accepting designs within the post-buckling regime. The third criterion is related to material failure in static load conditions, where the UL must be sustained under the *worst-case* conditions. This criterion is crucial to ensure that the structural design can withstand any unforeseen events that may occur during the aircraft's service life.

[4] As specified within CS 25.613 [4] where the appended paragraph AMC 25.613-4.3 reads: ..., *the effects of temperature and moisture ... should include the extremes of conditions...*

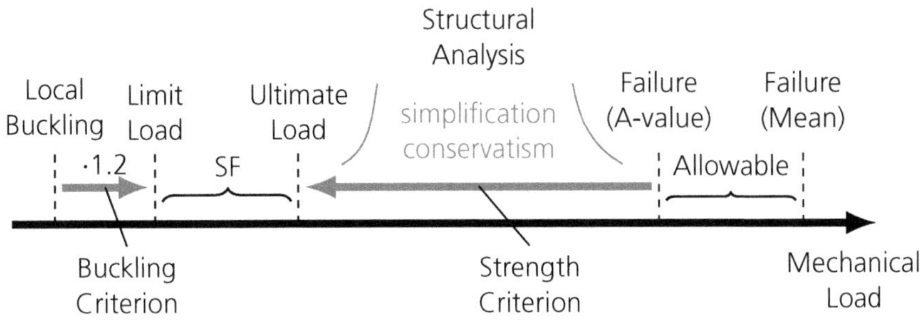

Fig. 2.3 Interpretation of conservatisms in static design of composite structures exposed to thermomechanical load conditions. Gray arrows indicate conservativity caused by structural analysis assumptions. The scaling is not representative. Adapted from Hachenberg [13]

The current methods used for sizing composite structures under thermomechanical conditions are based on simplified engineering approaches. These methods mostly rely on worst-case scenarios and multiple conservatisms to ensure a robust design. Figure 2.3 illustrates the character of these conservatisms over mechanical load. Local buckling, as stability design criterion, and the design loads are located on the left. Material failure, as another design criterion, is located on the right side highlighting a certain margin of safety resulting to the allowable definition. Structural analysis is conducted to verify the design criteria, see Fig. 2.1. These structural analyses use specific simplifications that must be defined conservatively to ensure a robust design. The conservativity introduced by structural analysis is indicated in Fig. 2.3 by gray arrows. For structures under thermomechanical load, a uniform temperature distribution is considered, applying the maximum service temperature as the worst-case condition. While this simplification leads to effective engineering-like sizing processes and simple analysis procedures, it is likely overly conservative. Figure 2.3 shows the resulting discrepancy between analysis results and real behaviour, illustrated in an exaggerated manner, using the strength criterion as an example. The discpreancy can be seen as additional MS or as overconservatism which ends up in unexploited loading capacity. However, the actual behavior of the structure can be captured by using a more detailed approach of temperature consideration. This allows to improev the utilization of structural load-carrying potential, or, with reference to Fig. 2.3 for illustration purposes, make the arrows shorter.

Design Challenges Related to Thermomechanical Load Conditions

The reasons for using this simplification is comprehensible from several points of economical considerations, experimental effort, or risk-assessment within the overall aircraft design:

- **Challenge 1: Design Loads:**
  The prediction of spatial temperature distributions and their variation over time throughout an aircraft's life is a complex and challenging task. Therefore, simplifying to generic worst-case conditions is understandable.

- **Challenge 2: Design Allowables:**
  Obtaining reliable material allowables at elevated temperature requires a high experimental effort[5]. This effort increases significantly when considering temperature distributions, which requires characterization at additional temperatures. From an economic perspective, the cost of such experimental effort could be unreasonable for specific aircraft development programs.
- **Challenge 3: Structural Analysis:**
  Accounting for realistic, non-uniform temperature distribution and its potential variation adds significant complexity to the analysis procedure. Hence, it is understandable that the current methodology simplifies the thermal conditions to uniform distribution, which is assumed to represent the worst-case scenario.
- **Challenge 4: Validation and experimental proof:**
  Verification and validation of analysis results and proof of structural strength is mandatory for certification. Structural testing complicates significantly if temperature variation is respected [15]. Although some test procedures have been published, e.g. by Rolfes et al. [16] or Liebisch et al. [17], such test programs are still associated with a high level of effort and uncertainties.

This work aims to contribute to these points by introducing an efficient methodology for deriving temperature-dependent material models and by considering spatial temperature distributions for structural analysis. The steps involved in achieving this objective are explained in detail in the following sections.

2.2 Objectives and Hypotheses

When exposed to external heating, real world structures develop a certain spatial temperature distribution. In contrast to that, current structural analysis strategies typically used for sizing apply a uniform temperature condition as simplification using the most severe thermal case.

Figure 2.4 provides a comparison between the spatial temperature distribution of such uniform thermal condition and two examples showing realistic temperature shapes. In both cases, example A and example B, non-uniform external heat sources lead to arbitrary temperature shapes and the hotspot temperature, i.e. the maximum temperature T^{max} is commonly used to assess the thermal behavior in relation to the allowable temperature of the material T^{alw}. Hence, this temperature is used in current uniform temperature analysis Uniform temperature Analysis procedure (UnA) (UnA) as shown in Fig. 2.4 left. Example A highlights the realistic temperature distribution related to this analysis case with the hotspot temperature T^{max} being equal to T^{alw}. However, this temperature is only reached in a small spot of example A and the majority of the part has a lower temperature. It can be concluded, that the average temperature is below T^{alw}.

[5] Other factors must also be considered, e.g. long-term studies regarding aging and fatigue, moisture uptake and related deterioration etc.

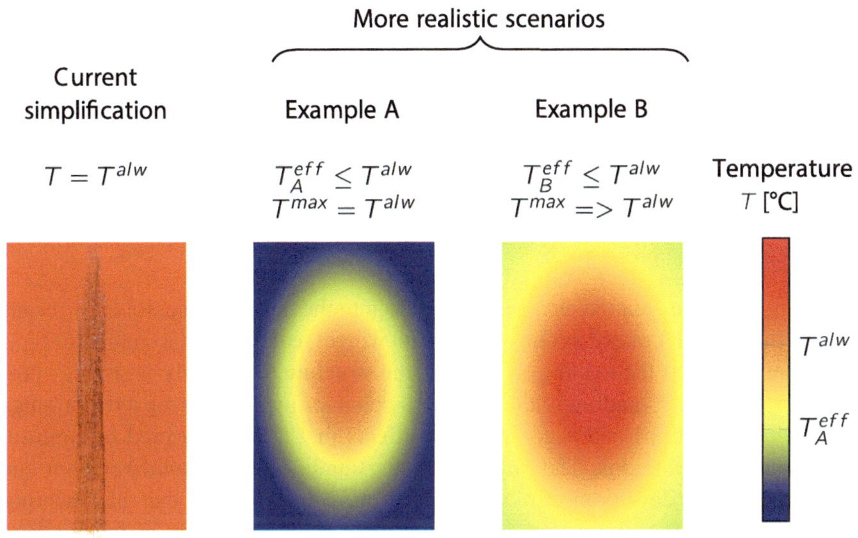

Fig. 2.4 Comparison of the spatial temperature distribution between the current analysis approach and realistic conditions

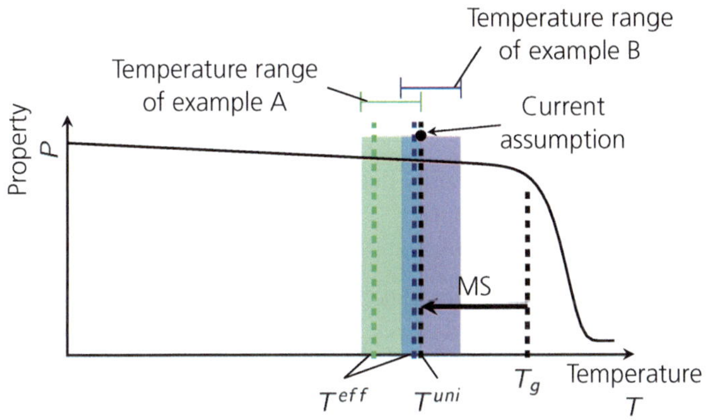

Fig. 2.5 Comparison between temperature ranges prevalent in realistic conditions and the assumptions made in the current analysis, with a focus on resulting mechanical material properties

Figure 2.5 provides a schematic illustration of the elastic material behavior over temperature for a material that corresponds to Fig. 2.4. The allowable temperature is a conservative value that is determined based on the glass transition temperature T_g using a certain MS, as described earlier. As mentioned, it is equal to the temperature T^{uni} applied in UnA and the black dashed line therefore marks the material property applied in current analysis. In comparison to that, the green area represents the temperature range of example A with the upper border being equal to T^{uni}. The green dashed line represents the temperature of the effective property in accordance with the average temperature. This leads to defining the central research hypothesis of this work:

Research hypothesis
An enhanced analysis procedure that considers spatial temperature distributions leads to a more efficient design of composite structures compared to current analysis procedures.

Integrating a more detailed analysis procedure into the design process, i.e. by means by distributed temperature, promises potential weight reduction, as it enables designers to better exploit the loading capacity of the structure. This analysis strategy provides a better understanding of the structural behavior by considering a wider range of thermal conditions and accounting for effects that are not covered by uniform temperature distributions. The improved understanding of structural behavior can further lead to the development of more advanced design concepts that may incorporate the use of composite materials in higher temperature domains, as indicated by the temperature distribution of example B in Fig. 2.4 and the according temperature range highlighted as blue area in Fig. 2.5. In this example, the effective temperature shown by the blue dashed line is shifted towards the allowable temperature of the material While the maximum temperature exceeds this limit.

For employing thermomechanical structural analysis under distributed temperature conditions in a sizing process, an enhanced analysis procedure has to consider following aspects:

- Elaborate modeling of temperature dependent material properties.
- Suitable and robust design criteria
- Consideration of thermal load variation to account for related uncertainty

Four working hypotheses are formulated as framework for developing the needed methods and investigating related topics. They are explained in further detail below.

In uniform conditions, it is only necessary to know the material properties at a single temperature. However, a comprehensive understanding of how the material properties change over a range of temperatures is important for dealing with distributed thermal conditions. This requires a significant amount of material characterization effort, which can be a particular challenge, as discussed earlier. Additionally, it is important to address the types of conservatism that are typically applied to material properties to ensure that the resulting formulations are compliant with regulatory requirements. This leads to the first working hypothesis:

Working Hypothesis 1:
The mechanical properties of Fibre Reinforced Plastic (FRP) composite materials can be modelled as function of temperature. The modelling leads to model errors in the properties, which are small compared to the uncertainty of the properties. These errors can be taken into account by conservative material allowable formulations.

In distributed thermal conditions, structural analysis must account for additional structural-mechanical effects compared to uniform temperature, for instance deflections normal to the composite plane. Understanding these effects and their sensitivities is essential for exploiting the load-carrying capacity of composite structures. Therefore, the second and third working hypotheses focused on investigating the suitability of methods for assessing the structural behavior (i.e., design criteria) and identifying the underlying mechanisms that contribute to the existing load-carrying potentials.

> **Working Hypothesis 2:**
> Typical methods used to assess the design criteria (stability, failure) are robust when used for structural analysis under distributed thermomechanical conditions.

> **Working Hypothesis 3:**
> Composite structures experience comparable mechanical behavior in uniform and distributed thermal conditions. In comparison, the consideration of distributed temperature with equal maximum temperature leads to higher critical loads (buckling, failure). This assertion is to be proven for simple examples of flat plates.

The sizing of structures is typically carried out through a deterministic analysis procedure. While the development of probabilistic structural analysis is state of the art in many fields and continuously advancing, industrial processes generally employ well-established procedures where the parameter uncertainties are accounted for by different MS. As noted earlier, the thermal conditions experienced by composite structures can be uncertain, resulting in variabilities of the temperature profiles and mechanical behavior. However, these uncertainties are not yet considered by a probabilistic approach. To address this challenge and enable a robust design, it is essential to consider these variations in thermal loading which is the focus of the fourth working hypothesis.

> **Working Hypothesis 4:**
> The mechanical behavior related to the occurring range of temperature variations of a structure can be captured by phenomenologically based structural allowables. Hence, these allowables contain the worst case conditions. This assertion is to be proven for simple examples of flat plates.

This fourth and final working hypothesis builds upon the previous working hypotheses. It consists of studies carried out on flat plates which serve to prove the research hypothesis by comparing the analysis results to the deterministic results under uniform conditions. This thesis is structured in a way to describe the related distributed temperature analysis DiA strategy and is presented in the following section.

2.3 Outline

Figure 2.6 provides an overview of the analysis methodology developed in this work. Each color corresponds to a specific topic of analysis. The numbers in circles indicate the section where the topic is described. The methodology is a systematic approach which has the potential to be employed for varying applications. The chapters and sections that correspond to the individual topics aim to provide a clear overview of the input, method description and output. By following this procedure and in line with the objective of this thesis, structural designs with improved performance over temperature or lower weight could be achieved.

The red blocks represent input parameters required at the beginning of the design process. They include the structural setup as well as thermal and mechanical boundary conditions. Chapter 3 discusses the material characteirzation, which is of particular relevance to this work. This chapter explains the methodology and results of a material characterization program for a selected Carbon Fibre Reinforced Plastic (CFRP) material at various temperatures. The material characterization results are used as input for material modelling, which is explained in Chap. 4. The objective here is to establish a functional relationship between material properties and temperature. In accordance with working hypotheses 1, the material models derived represent allowable properties that include inherent uncertainties and model errors. Both, the material characterization and material modeling tasks are typically carried out before or in parallel with the early design stage, as they provide the essential framework for structural analysis

Chapter 5 explains the individual steps involved in the structural analysis process, indicated by the dashed line in Fig. 2.6. A deterministic structural analysis procedure is highlighted by the gray area. It is used to determine the structural behavior at a specific temperature distribution. It has to be emphasized, that this part of the analysis procedure is applied similarly for both, UnA and DiA. These deterministic analyses are built on FEA consisting of multiple steps shown in yellow boxes. They provide the input for the blue boxes which represent the assessment procedures to evaluate the design criteria. Results obtained at different temperature distributions, e.g. by multiple repetitions of deterministic analyses, are then used in a probabilistic analysis, resulting in a statistical distribution of the design criteria. Based on this, structural allowables are derived that are representative of the structural behavior over the range of applied thermal conditions. This analysis procedure presents an alternative to the common procedure, where uniform thermal conditions are considered through a single realisation of the deterministic structural analysis chain.

Chapter 6 utilizes the analysis approach explained in Chap. 5 and the material models derived in Chap. 4 to compare the common uniform temperature approach with the novel approach of using distributed temperatures. These analyses and comparisons are carried out using a simple example of a flat plate and applying several variations of temperature distribution. This allows for a comprehensive discussion of working hypotheses 2–4 and highlights the differences between the UnA and DiA analysis methodologies. Finally, a summary of the work is provided in Chap. 7 and the conclusions are evaluated with respect to the research objective.

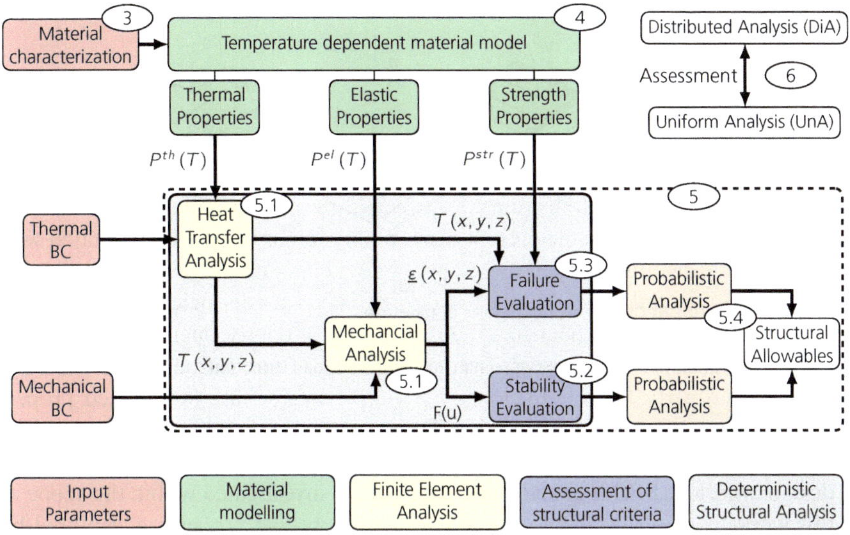

Fig. 2.6 Overview to the content of this thesis shown by a schematic overview to the overall analysis strategy. Chapters related to an individual topic are highlighted by numbers in circles

2.4 Notations and Scope

This section presents an overview to specific denotations and conventions used throughout the work.

The term *material properties* is used as a collective representation for all elastic and strength properties as well as for thermal expansion coefficients. In this general form, it is denoted by the symbol P. To distinguish between different load directions, the index notation P_{ijk} is adopted. In reference to Fig. 2.7 the subscripts i and j refer to load directions related to the coordinate system. The index k is used to distinguish between tensile and compressive loads, respectively. Stress and strain quantities use positive and negative magnitudes to express tension and compression loading and therefore do not use the index k. The lamina's coordinate system employs numerals (123), e.g. for a single Unidirectional (UD) ply, whereas the laminate is designated by lettered coordinates (xyz).

Following terms and abbreviations are used to distinguish between the analysis approaches:

- The common approach: UnA—represents a single deterministic analysis in uniform thermal conditions.
- The novel approach: DiA—represents either the deterministic analysis in distributed thermal conditions or the extensive analysis procedure using a set of these analyses that additionally considers variations of thermal loads.

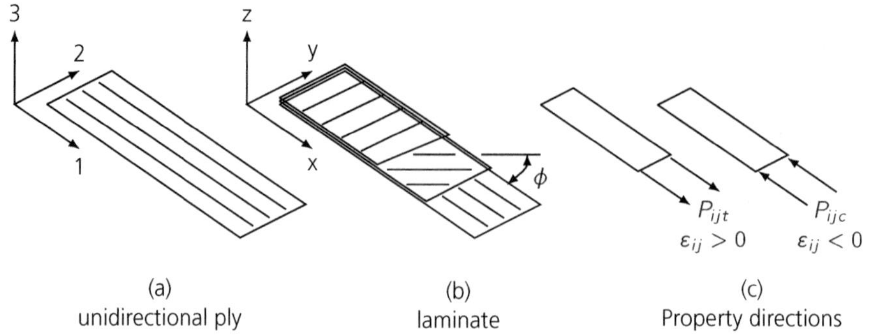

Fig. 2.7 Index notations indicating coordinate systems and load directions

- Design values refer to the mechanical loads that are associated with the respective design criterion. Buckling and static strength are investigated within the scope of this work.

The following conditions are applied:

- Material properties are considered by their dry and unaged margins.
- The thermal response is considered by stationary state temperature distributions.
- Mechanical loads are considered by static conditions.

References

1. E. Torenbeek, *Synthesis of Subsonic Airplane Design*. (Springer Dordrecht, 1986). https://doi.org/10.1007/978-94-017-3202-4
2. D.P. Raymer, *Aircraft Design: A Conceptual Approach: Fourth Edition*, 4th edn. AIAA education series. (Cambridge University Press, 2006). 1-56347-829-3
3. European Aviation Safety Agency, Certification Specifications for Normal-Category Aeroplanes (CS-23): Amendment 5 (2017)
4. European Aviation Safety Agency, Certification Specifications for Large Aeroplanes (CS-25): Amendment 3 (2007)
5. European Aviation Safety Agency, Certification Specifications, Acceptable Means of Compliance and Guidance Material for Small Rotorcraft (CS-27): Amendment 9 (2021)
6. European Aviation Safety Agency, Certification Specifications, Acceptable Means of Compliance and Guidance Material for Large Rotorcraft (CS-29): Amendment 10 (2021)
7. European Aviation Safety Agency, AMC-20: General Acceptable Means of Compliance for Airworthiness of Products, Parts and Appliances: Amendment 20. 20 (2020). https://www.easa.europa.eu/document-library/certification-specifications/amc-20-amendment-20
8. M.W. Long, J.D. Narciso, Probabilistic Design Methodology for Composite Aircraft Structures. Report: DOT/FAA/AR-99/2. (1999). https://apps.dtic.mil/sti/tr/pdf/ADA365683.pdf
9. M. Niu, Airframe Stress Analysis and Sizing. 2nd ed. (Hongkong Conmilit Press Ltd., 1999). ISBN: 962-7128-08-2. https://www.academia.edu/32132236/AIRFRAME_STRESS_ANALYSIS_AND_SIZING_BY_MICHAEL_C_Y_NIU

10. C.C. Chamis, R.F. Lark, J.H. Sinclair, Integrated Theory for Predicting the Hygrothermomechanical Response of Advanced Composite Structural Components. Advanced Composite Materials - Environmental Effects (1978), pp. 160–192. https://doi.org/10.1520/STP34863S.
11. T.A. Collings, R.J. Harvey, A.W. Dalziel, The use of elevated temperature in the structural testing of FRP components for simulating the effects of hot and wet environmental exposure. Composites **24**(8), 625–634 (1993). https://doi.org/10.1016/0010-4361(93)90125-R
12. A. Revathi et al., Effect of hot-wet conditioning on the mechanical and thermal properties of im7/8552 carbon fiber composite. Indian J. Adv. Chem. Sci. **2**, 84–88 (2014). https://www.researchgate.net/publication/263922566_Effect_of_Hot-Wet_Conditioning_on_the_Mechanical_and_Thermal_Properties_of_IM7_8552_Carbon_Fiber_Composite
13. D. Hachenberg, *The Role of Advanced Numerical Methods in the Design and Certification of Future Composite Aircraft Structures* (World congress on computational mechanics, WCCM V, 2002)
14. R. Degenhardt et al., COCOMAT-improved material exploitation of composite airframe structures by accurate simulation of postbuckling and collapse. Compos. Struct. **73**(2), 175–178 (2006). https://doi.org/10.1016/j.compstruct.2005.11.042
15. J. Singer, J. Arbocz, T. Weller, *Buckling Experiments: Experimental Methods in Buckling of Thin-Walled Structures: Basic Concepts, Columns, Beams and Plates* (Wiley, Chichester and New York, 2002). 9780471956617. https://doi.org/10.1002/9780470172988
16. R. Rolfes, S. Tacke, R. Zimmermann, Development and experimental verification of FE-model for stringer-stiffened fibre composite panels under combined thermal and mechanical loading conditions, in *European Conference on Spacecraft Structures, Materials and Mechanical Testing, Braunschweig* (Nov. 1998), pp. 109–115. https://www.dlr.de/fa/Portaldata/17/Resources/dokumente/institut/1999/s1999
17. M. Liebisch et al., Thermomechanical testing of CFRP structures under varying thermal conditions, in *SAMPE EUROPE Conference and Exhibition 2021*, Baden, Switzerland (2021)

Chapter 3
Characterization of Composite Properties at Different Temperatures

3.1 Scope of This Chapter

A temperature-dependent formulation of material properties is essential to analyze the structural behavior under distributed thermal conditions, which is covered by working hypothesis 1 (see Sect. 2.3). This chapter describes the experimental characterization program and the evaluation procedure employed to determine material properties at a range of distinct temperatures. The focus is on the mechanical properties needed for analyzing the design criteria (buckling and strength), conf. Sect. 3.3. Thermal properties are not investigated by this characterization program. The resulting property-temperature data serves as input for advanced material modelling and are used to derive property functions of temperature, which is discussed in the subsequent chapter (Chap. 4, *Modelling of temperature-dependent composite properties*).

A comprehensive characterization program is essential to capturing the temperature-dependent properties of composite materials and to developing an understanding of their behavior under thermal loads. However, testing in specific temperature ranges can be challenging and complicated and the amount of testing required for such programs can become substantial. This context is identified as a relevant challenge in the scope of aircraft design programs, see Sect. 2.2. The characterization program described in this work aligns with industrial test standards and therefore enables estimation of the necessary test quantities needed in such development environments. In essence, the program may serve as starting point for others and provides a valuable reference for testing FRP composites at high temperatures. It is important to note that the characterization results are exclusive to the CFRP material discussed in Sect. 3.2, and are applicable only to this material.

Polymers typically show a distinct temperature-dependent behavior due to their molecular structure. Elevated temperatures cause a reduction of the macroscopic properties due to the softening of the polymer [1–3]. This characteristic is discussed in detail in Sect. 4.2.1. For the investigations throughout this thesis, the temperature range of 300 °C to 250 °C is used, which is relevant for many polymers. In contrast

M. Liebisch, *Design Load Allowables for Composite Plates Exposed to Thermomechanical Loads*, Mechanics and Adaptronics,
https://doi.org/10.1007/978-3-031-71352-1_3

to polymers, carbon fibers exhibit relatively low sensitivity to temperature changes within such temperature range [4, 5]. The orthotropic nature of composite materials, as for instance existing in a UD composite lamina, is a result of the fiber arrangement, while the fibers itself can also be orthotropic. However, the sensitivity of such orthotropy on the composite material properties can be estimated using homogenization rules and Classic Lamination Theory (CLT), which has been previously reported [6–8, 10]. The homogenization rules are based on parallel and series connections, which model the composite behavior parallel and perpendicular to the fiber, respectively. In fiber direction, the fibers dominate due to their high properties, while in the transverse direction, the matrix must transfer loads between the fibers, resulting in a greater impact of reduced matrix properties due to thermal sensitivity. The characterization program discussed within the current chapter provides a detailed discussion on the UD composite material's property response to temperature changes and its complex orthotropic behavior. This is expressed through the resulting lamina properties captured at a range of distinct temperatures.

The assumption of linear stress-strain behavior is commonly made for CFRP materials. However, this assumption may not hold in all cases, and nonlinearity can arise due to changing plasticity onset or other factors that vary over temperature [11]. Investigating such nonlinear behavior, including plasticity or viscoelasticity, typically requires elaborate test programs [12, 13]. However, such investigation is not the primary focus of this work and testing is conducted under quasi-static conditions and constant load rates. While the linearity of the stress-strain relations is verified to some extent, potential physical causes of nonlinear behavior are not further investigated in this study. This assumption creates a restriction to linear-elastic material models, which is considered throughout material modelling, see Chap. 4.

3.2 Material Definition

To design structures that can withstand thermomechanical loads, it is crucial to select a material that provides the necessary thermal resistance. Alternatively, a designer may need to improve the thermal design of the structural environment, e.g. by adding heat barriers or insulation which finally add weight, to meet the allowable temperature limits of a pre-selected composite material. Hence, the ultimate goal of related structural design optimization is to enable higher thermal loads while meeting mechanical load requirements and minimizing weight. In this context, a new composite material, designated as Toughened Cyanate Ester Carbon Fiber Composite Material (tCFRP), has been identified as a potential candidate for use in high-temperature conditions. Compared to conventional CFRP materials, this material offers improved performance at high temperatures and can withstand even higher maximum service temperatures. As such, it could be an excellent option for designing weight-optimized composite lightweight structures that are exposed to high thermomechanical load conditions [9, 14]. The present study benefits from this material's wider temperature range. The details of the material specifications are explained below.

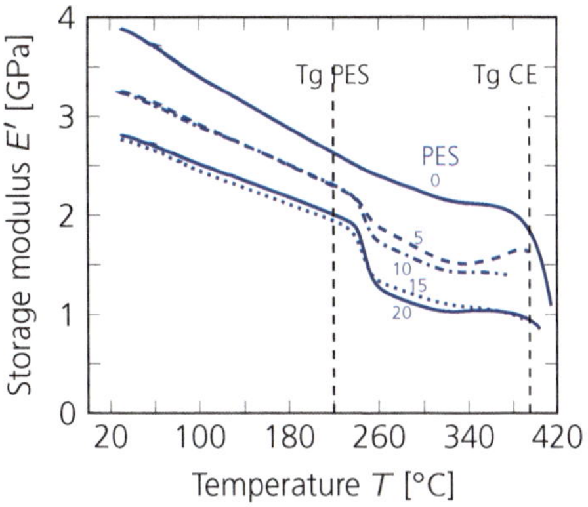

Fig. 3.1 Comparison of the neat Cyanate Ester to toughened Cyanate Ester versions with different quantities of PES. The blue numbers refer to the PES content in phr

The tCFRP material solution consists of the Cyanate Ester (CE) PT30 as the constituent matrix material that governs promising properties, such as high thermal stability, chemical resistance, low water absorption, and high strength [15]. However, the high crosslink density of neat CE results in inherent brittleness and a low fracture toughness [16], despite its high T_g of approximately 400 °C [17, 18]. To address the brittleness issue, fracture toughness can be improved through mixing with specific materials or by additive interface layers [19–22]. To achieve a less fragile material solution, the CE resin is mixed with PES[1] which results in varying material behavior, as shown in Fig. 3.1. Neat CE exhibits an almost linear decrease in storage modulus until reaching T_g while having a relatively high magnitude. However, the addition of PES causes a lower modulus, with a magnitude that approaches that of neat PES. Notably, a softening step around 230 °C is observed due to the PES glass transition at 220 °C [24] while a certain margin of elasticity is remained. Ultimately, the solution with 15phr[2] of PES2603MP as the additive to the PT30 CE resin is selected. As shown in [24], this type of PES produces lower drop of tCE storage modulus compared to others. The selected quantity of PES represents a compromise between reducing storage modulus and increasing in fracture toughness, while the tCE resin material exhibits approximately double values compared to the neat CE matrix [24]. Table 3.1 presents selected properties of the composite constituents, highlighting the difference between the matrix constituents PT30 and PES. Notably, the ultimate elongation of the fiber value is higher than that of PT30. Amirova [24] compared the neat PT30 material to the tCE under Room Temperature (RT) conditions, where toughening caused a reduction of the flexural modulus from 3802 Mpa to 3299 Mpa, while the flexural strength increased from 45.3 Mpa to 54.6 Mpa. Unfortunately, there is no additional information available yet regarding other matrix properties.

[1] Properties of PES are exemplarily obtained by Amsler and Frey AG [23].

[2] Parts per hundred rubber.

Table 3.1 Properties of the constituent materials for the tCFRP material at RT conditions.

Constituent	Density	Young's modulus	CTE	Tensile Strength	Ult. elongation
	ϱ $[\frac{g}{cm^3}]$	E_t [GPa]	α_{11} $[\frac{1}{K}]$	R_{11_t} [MPa]	$\varepsilon_{11,ult}$ [%]
T800-S [25]	1.8	294	$-0.4 \cdot 10^{-6}$	5880	2.0
PT30 [17]	1.2	4.07	$28 \cdot 10^{-6}$	41	1.2 to 1.5
PES [23]	1.37	2.7	$55 \cdot 10^{6}$	90	15

The tCFRP material discussed within this study is a UD thin ply composite material which has a nominal ply thickness of $t^{ply} = 0.06$mm. It was manufactured by *North Thin Ply Technology* as prepreg with a fiber areal weight of $m^{ply} = 67$g/m^2 and fiber volume ratio of $V_f = 58\%$. The composite material consists of *T800-S* carbon fibers and the tCE matrix material discussed above. The prepreg material is slit and prepared for automated fiber placement (AFP) preforming, and is cured using a typical prepreg vacuum bagging autoclave process. The cure process is divided into two stages, with temperature levels of 120 °C and 180 °C for 120 min each, see Fig. 3.2. The heating rate to reach both temperature levels is 2 K/min, while the cooldown to RT is controlled at 1 °C/min. After curing, a postcure process is required to fully cure the matrix and to achieve the intended high T_g. The postcuring is done in a separate oven and without the autoclave bagging. The temperature is raised to 180 °C at a heating rate of 2 K/min, and then to 260 °C at a lower heating rate of 1 K/min, producing an almost fully cured resin. The cooldown is again performed at a controlled rate of 1 K/min. It is important to note that viscoelastic relaxation and thermal stress development, i.e. residual stress from curing, are critical factors for failure onset, and these are discussed in detail in Sect. 5.1.4.

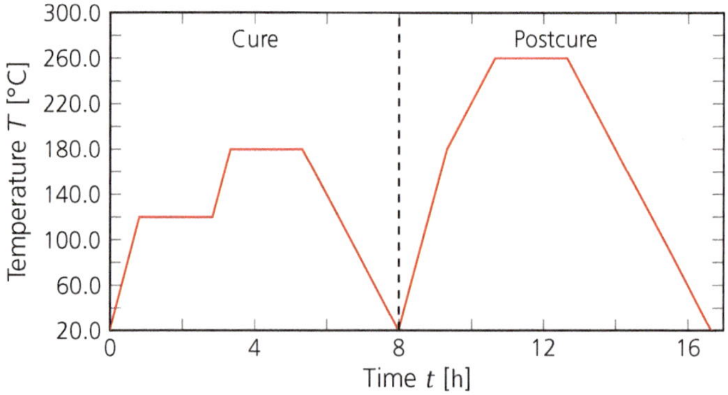

Fig. 3.2 Cure cycle of the tCFRP material solution

3.3 Overview to the Characterization Program

The focus of the present study is on thin-walled composite structures, which are typically analyzed under the conditions of plane stress [10]. Therefore, the material characterization mainly focuses determining the in-plane properties of the UD material. To analyze the thermomechanical behavior of the structure, it is necessary to know the elastic properties, strength properties, and thermal expansion characteristics with respect to the orthotropic behavior of the material. Material properties can show different magnitude in tensile and compressive load conditions and they can vary significantly in their relation to temperature. Hence, both, tensile and compressive load conditions have to be considered by the characterization program to generate a suitable knowledge of the material behavior. The material softening process introduced by the glass transition of the PES toughener represents a crucial temperature domain. The motivation of this work indicates a potential use above T_g for small regions of a structure. Therefore, the related material property degradation has to be captured and the temperature range of interest in this study is define up to 250 °C.

The selection of appropriate tests prior to starting a material characterization program is crucial, as it involves committing to specimen dimensions and test conditions. However, this can be a challenging task due to the sensitivity of test fixtures to specimen behavior at varying thermal conditions, the availability and maximum operating temperature of fixtures, and the complexity of handling and aligning fixtures at high temperatures. While Carlsson and Pipes [26] provide an overview of standard methods, they do not provide insights into test procedures, handling, or sensitivities at elevated temperatures. Moreover, both ASTM[3] and European standards (EN) lack sufficient guidance on elevated temperature testing. Therefore, initial experiences have been developed by the author [27–29] by using available fixtures and applying standard conditions based on best practices. The test standard selection shown below is based on these experiences and related details are explained within the respective sections of result discussion (see Sect. 3.4). These sections further provide an overview to relevant literature on the test procedure and the respective property behavior. All tests are performed inside a thermal chamber that is constructed around the load adapters. An overview of the specimen dimensions and principle of the test fixtures can be found in the Appendix (see Table A.1 and Fig. A.1). Details in the preparation and testing procedures employed are explained in the following sections.

- Tension: ASTM D3039 [30]
- Compression: EN 2850 [31]
- In-plane Shear (IPS): ASTM D3518 [32]
- Interlaminar Shear Strength (ILSS): DIN EN 2563 [33]

[3] ASTM International, international standards organization, formerly known as American Society of testing and Materials.

- Coefficient of Thermal Expansion (CTE) determined wihin thermogravimetric analysis (TGA) setup, no standard adressed
- Dynamic Mechanical Analysis (DMA) in bending mode, no standard adressed.

3.3.1 Specimen Preparation

Flat UD plates and plates are preformed to the required laminate thickness. Subsequently, the curing process is achieved through vacuum bagging, following the procedure described in Sect. 3.2. The UD plates are used for specimen (tension, compression, ILSS) and the ±45° plates are used for the IPS specimen as well as for tabs. Ultrasonic scans are made to verify the quality of the plate material. Typically, tabs are made of standard ±45° glass fiber FRP material and bonded to the specimen at RT. However, with respect to the test temperature range, this material and typical tab adhesive do not provide the required performance. To overcome this problem, tabs are made of equal thickness tCFRP material, while a film adhesive made of the identical tCE matrix material is used for bonding the tab and specimen. Fully cured tab plates and specimen plates are then bonded in a heated press, following the materials curing cycle followed by a freestanding post-cure, as shown in Fig. 3.2. After tab bonding, specimens are cut out of the plates following the standard dimensions and each specimen dimension is finally quantified to ensure the geometrical tolerances required for testing. Strain gauges of the HBM KFU series are then adhered using the polyimide adhesive system PI-32, which is capable of withstanding the indicated temperature range [28]. The HBM KFU strain gauges can operate at temperatures up to 350 °C [34] and are delivered with attached wires in a three-wire technique that avoids difficult and error-prone soldering with high-temperature-resistant solder. This is an experience-based recommendation for planning tests at such temperatures. The wires are packed into a mesh to simplify bonding. Since the PI-32 bonding agent is a viscous liquid at RT, strain gauges are aligned using neoprene pads and metal clamps to apply the required contact pressure between the gauge and specimen. Based on experiences, it has to be verified that the metal clamps produce the needed clamping force at adhesive cure conditions, as metal springs relax at elevated temperature. The adhesive is cured using a typical oven process following the specified 100 °C and 200 °C dwelling steps. A specific post-curing step, as specified by the manufacturer, is not conducted to avoid possible thermal degradation effects that may occur within any of the attached parts. DSC (Dynamic Scanning Calorimetry) investigations performed previously have shown a sufficient quick increase in the T_g of the PI-32 material system during the preheating and dwelling process of the test procedure itself, ensuring the required performance of the adhesive. Initial trials have confirmed the sufficient applicability of this procedure.

3.3.2 *Test Procedure*

To ensure accurate test results, all static tests are conducted within a thermal chamber that fully covers clamps and fixtures and therefore the clamping devices are also heated. Related sensitivities on the test results have been studied prior to the test program [35] for selecting a suitable test procedure. When specimen are aligned within the chamber, the clamps act as heat-sinks due to their rigid steel construction, resulting in a temperature profile along the specimen with the lowest temperature at its clamped ends and the highest temperature at the center between the clamps. This is in line with the test objective, as the area of interest is within the center of the specimen. However, to produce similar conditions for all specimen, the clamps are heated at test temperature for a dwell time of one hour before starting to test. As the rigid steel clamps remain a certain heat sink capability, this results in temperatures between 20 °C and 30 °C below the target temperature, measured at the clamping surface. Thermal contact produced by the clamping ensures a similar temperature of the clamps and the specimen within the mounting areas. In contrast, the specimen center location reaches the target temperature with good accuracy and reproducibility within a few minutes in the free length, ensuring temperature tolerances of 3K and ±5K for $T < 200\,°C$ and $T > 250\,°C$ respectively, as required by the EN 2744 standard [33]. This leads to a temperature gradient between the specimen center and clamps. However, other thermomechanical test setups, such as those using local heating devices at the free length of the specimen, as demonstrated by Li et al. [5] and Zhou et al. [36], exhibit much higher temperature gradients along the length of the specimen. The sensitivity of these gradients to the resulting properties has not been studied in detail. The central location of the specimen, where the target temperature is achieved, is the primary domain of interest for the test and defines the location of measurement and intended failure location. The individual test procedures used for each setup align with temperature specification as defined by the EN 2744 [37] standard and are described below.

All tests using the tensile clamping setup (tension and IPS) apply an initial dwell time of 60 min prior to testing, as explained above. In addition, testing of each specimen is done using a two-step dwell process. In the first step, the specimen is aligned at the top side only and dwelled for a minimum of 5 min to allow thermal expansion. This period is selected and confirmed by strain gauge signals approaching a constant level. After opening the chamber to fix the bottom clamps, another 5 min dwell time is applied. Subsequently, the tests are performed according to their standard specification.

In compressive testing, the test machine is equipped with two horizontal and parallel load adapter plates, which are finally used to introduce the loads into the specimen. The used test fixture, as shown in Fig. 3.7, represents a small assembly that is aligned in between these load adapter plates and can be removed from the thermal chamber for mounting the specimen. This unique feature requires slight adaptations to the standard procedure. Since the free length is relatively small and almost the entire specimen is connected to the rigid test fixture, a fixture dwell time of at least 60 min is necessary, using a target test temperature of 3 K above the intended test temperature. The fixture used is less rigid compared to the tensile setup, and therefore heats through more quickly reaching a suitable state by the described procedure. This procedure is based on initial experiences and following this process generates the intended test conditions at the free length region. Specimen alignment is conducted manually after taking the fixture out of the chamber. To reduce influences from thermal expansion, a two-step dwelling process similar to the tension setup is conducted. In the first step, the fixture screws are tightened only enough to keep the specimen barely aligned while there is no connection between specimen and adapter plates. Before the second step, the clamps are tightened according to the standard specification. Both steps involve a 5 min dwell time. To confirm good agreement with the intended test temperatures, thermocouples with high-temperature resistant sticky tape are applied to the specimen side at the center location of the free length. After that, the specimen is connected to the load adapter plates via the test machine and testing is then according to the typical standard specification.

3.3.3 Evaluation Methodology

Modulus and Strength Evaluation

The individual material properties, as well as the stress and strain components, are defined using the notations for load directions introduced in Sect. 2.4.

The Young's modulus is determined by linearizing the relationship between stress and strain at two measurement points, represented by indices a and b. Hence, the equation for determining the Young's modulus is given by:

$$E_{ijk} = f(a, b) = \frac{\sigma_{b,ij} - \sigma_{a,ij}}{\varepsilon_{b,ij} - \varepsilon_{a,ij}} \quad \text{with} \quad |\sigma_{b,ij}| > |\sigma_{a,ij}| \quad \text{and} \quad |\varepsilon_{b,ij}| > |\varepsilon_{a,ij}| \tag{3.1}$$

The measurement points a and b correspond to specific test standards and load levels, which are provided in the relevant result discussion. In the case of IPS testing and compression testing, additional conversions and checks of the measured strain history are required and they are described in detailed in their respective sections. However, Eq. 3.1 is applied in a similar manner.

The strength components represent the maximum load the UD material can withstand in unilateral loading conditions. These components assume a homogeneous state of stress within the specimen, and hence, smear the heterogeneous behavior of the fiber and matrix. In the scope of characterization testing and respecting the load directions, the strength value is obtained by determining the maximum stress the material can withstand:

$$R_{ijk} = max\left(|\sigma_{ij}|\right). \tag{3.2}$$

The aforementioned procedure is in line with the standards. The specimen fractures have to be assessed regarding fracture type and failure location to verify the test validity, which is also in line with standards. Further details are explained within the respective sections.

An alternative method for determining the strength criterion is the linear offset methodology. This involves using the Young's modulus and Eq. 3.1 to draw a linear slope. An offset line is constructed by shifting this slope through a defined strain offset. The strength value is obtained by the intersection point of the stress-strain curve and the offset slope. This methodology can be useful if nonlinear stress-strain relations are present.

$$R_{ijk}^{off} \rightarrow |\sigma_{ijk}| < \left(|\varepsilon_{ijk}| - \Delta\varepsilon^{off}\right) \cdot E_{ij} \tag{3.3}$$

Statistical Evaluation

Statistical evaluation is conducted using common methods and based on a number of n test repetitions, each providing an individual value of the property $sym : prop$. The set of these property values is then used for statistical analysis. Assuming a normal distribution, the mean value of the property P is determined by:

$$\mu_P = \frac{1}{n} \sum_{i=1}^{n} P_i \tag{3.4}$$

The width of the scatter can be described by either the variance var_P or the standard deviation s_P of the property P.

$$var_P = \sum_{i}^{n} (\mu_P - P_i)^2 \qquad s_P = \sqrt{var_P} \tag{3.5}$$

The variation coefficient cv is defined as the ratio between the standard deviation and the mean value of the property P. This quantity is useful for comparing the variability of properties, and is especially important when comparing property degradation over temperature.

$$cv_P = \frac{s_P}{\mu_P} \cdot 100 \tag{3.6}$$

The evaluation of these statistical quantities provides information on the scatter behavior of properties. Statistical analysis is an important part of characterization programs and is required for deriving design allowables. However, due to the limited number of test repetitions, the statistical representativity of the present program is low and therefore the results obtained serve only as rough estimates.

Linearity Assessment

Preventing permanent deformation is a key design criterion at LL conditions. Damages or onset of plasticity are often associated with a nonlinear stress-strain relation.

Henry et al. [38] investigated tensile test methods for determining the onset of plasticity and found the offset-strength criterion, see Eq. 3.3, to be effective in identifying the yield onset. The offset strain for this method is defined as $\Delta\varepsilon^{off} = 0.2\%$. A similar value is proposed in IPS testing (conf. ASTM D 3518 [32]). Hobbiebrunken et al. [39, 40] used an offset coefficient of $\Delta\varepsilon^{off} = 0.05\%$ for transverse tensile testing of FRP materials. In all of these cases, the linear offset strain $\varepsilon_{ij}^{lin,off}$ is obtained by Hooke's law and the offset coefficient $\Delta\varepsilon^{off}$ according to Eq. 3.7. The intersection of linear offset strain to the measurement curve designates the point of nonlinearity onset. $\Delta\varepsilon^{off}$ represents a scale for the criticality since it is sensitive to this point of nonlinearity onset. The value used by Hobbybrunken et al. [39, 40] seems to represent a relatively conservative margin leading to low stress values. Henry et al. [38] indicate, that the magnitude of strain-offset is a value of experience. To determine the onset of irreversible effects, they propose an alternate test methodology using step-wise loading-unloading cycles to record strain deviations at unloaded state. However, this methodology is sensitive to the resulting failure loads and therefore would require additional specimen for determining the nonlinearity onset.

In addition to the onset of irreversible damages, linear elasticity is a basic assumption of structural analysis. Hence, a methodology to assess the linearity of the stress-strain relation is defined in the following for verifying the assumption of linear elasticity.

$$\varepsilon_{ij}^{lin,off}(\sigma) = \frac{\sigma_{ij}^{act}}{E_{ijk}} + \Delta\varepsilon_{ij}^{off} \tag{3.7}$$

This relation describes the linear offset strain which can be compared to the actual measurement data and refers to the description above. Normalizing the difference of actual strain and linear offset strain by the strain offset magnitude leads to the linearity ratio r^{lin} as a more practical quantity for assessment:

$$r^{lin} = \frac{\varepsilon_{lin,off}(\sigma) - \varepsilon_{act}(\sigma)}{\Delta_{off}} \tag{3.8}$$

Because of normalization, the following domains can be classified as representative in terms of linearity:

- $r^{lin} > 2$: progressive stress-strain behavior designated as unacceptable nonlinearity
- $2 > r^{lin} > 1$: progressive stress-strain behavior within acceptable extent
- $r^{lin} = 1$: ideal linear elasticity
- $0 < r^{lin} < 1$: degressive stress-strain behavior within acceptable extent
- $r^{lin} < 0$: degressive stress-strain behavior designated as unacceptable nonlinearity.

Within the scope of this work, nonlinearity is evaluated based on the criterion explained above through assessing the trend of r^{lin} at different strain levels over temperature. To this, the offset strain of $\Delta\varepsilon^{off} = 0.2\%$ is employed. However, this evaluation depends on the margin of $\Delta\varepsilon^{off}$. Therefore, the procedure has to be treated as a tool for classifying the stress-strain relation into linear or nonlinear despite using it as a valid tool for estimating nonlinearity onset. Nevertheless, it is integrated into the present evaluation procedure as nonlinear elasticity is an important factor at elevated temperature. Efficient and accurate methods for determining the onset of nonlinearity remain part of research and the present work helps to get a better understanding on it.

3.4 Material Characterization Results

This section presents the material characterization results for the individual properties and tests carried out. Each section describes the related test and evaluation conditions and the observations are discussed for each property in detail focusing on elastic behavior, strength behavior and the relation to temperature. Additionally, a comparison between the properties and a comprehensive discussion of the general material behavior can be found within the following Sect. 3.5.

3.4.1 Dynamic Mechanical Analysis

Figure 3.1 compares tCE versions with different amount of PES toughener material and provides valuable insights into the effect of the PES on the behavior of the tCE. However, the tCFRP composite material additionally includes carbon fibers which are sensitive to the material behavior. Hence, similar DMA measurements are conducted for the composite to capture the sensitivity of the matrix on the composite and its behavior over temperature. The measurements are carried out using a three-point-bending DMA setup, applying a constant force (at 1.0 Hz) while heating the specimen continuously at a rate of 2.0 $\frac{\mathrm{k}}{\mathrm{min}}$. The dynamic loading character of DMA is useful for several purposes, and the viscous part can be used for determining T_g (e.g. Li et al. [5]) or to study the materials viscoelastic relaxation behavior (e.g. Hein [41]). However, only the elastic parts are studied here.

Figure 3.3 presents the storage modulus of the UD composite over temperature in the fiber direction (left) and transverse direction (right). The measurement in the fiber direction reveals an almost constant modulus for temperatures below the T_g of PES at around 235 °C. Beyond this temperature, softening is observed, and the modulus decreases almost linearly between 280 °C and 340 °C . The specimen develops plastic deformation beyond this temperature range and is not further representative.

Two measurements are performed in the transverse direction (right chart), represented by solid and dashed lines. The storage modulus decreases linearly until PES T_g, similar to the neat matrix behavior. Temperatures above PES T_g lead to degressive trends, and irreversible specimen deformation occurs above 300 °C. This is beyond the temperature domain of interest and is therefore not further investigated. However, this observation is similar for both, specimen in fiber direction and specimen in transverse direction. The different temperature relation highlights the

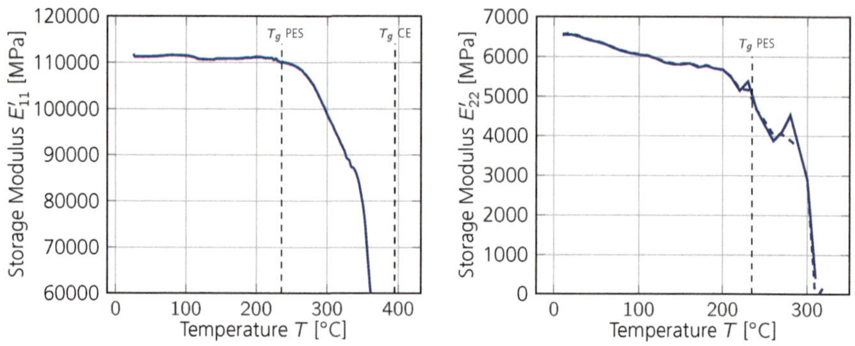

Fig. 3.3 Storage modulus (blue) measured by DMA in fiber direction (left) and transverse direction (right). Vertical dashed line indicate softening as they represent the glass transition temperatures of the matrix components

dominating characters of fiber dominance and matrix sensitivity between the different fiber direction setups. Notably, the low forces applied in DMA do not provide full comparability to the static measurement results.

3.4.2 *Tensile Loading*

Test Procedure

Tension testing is performed following the guidelines of ASTM3039 [30]. The specimens are mounted at the top and bottom using a wedge mechanism clamping device, as shown in Fig. A.1. This device causes increasing clamping force at increasing tensile forces. Strain measurements are taken through strain gauges located at the center of the specimen. Half of the specimens are equipped with longitudinal strain gauges, while the other half have both longitudinal and transversal strain gauges. The latter are used to determine the Poisson's ratio, as described in Sect. 3.4.3. As shown in Table A.1, different specimen geometries are used for tensile loads in the fiber direction and in the transverse direction. They are common and in line with the standards while the thickness is selected in deviation to the recommendations. The thickness of the fiber direction specimen is reduced to produce lower failure forces, while the thickness of the transverse specimens is increased to have a less fragile specimen. In preliminary tests, ±45° tabs are used according to the tabbing method explained in Sect. 3.3.1, but this led to poor behavior, including premature failure, specimen slipping, or tab delamination. Therefore, the results presented here are based on specimens without tabs. The Young's modulus is determined in both the longitudinal direction ($E_{11,t}$) and transverse direction ($E_{22,t}$) following the ASTM D3039 recommendations, between $\varepsilon_a = 1000\,\mu\text{m/m}$ and $\varepsilon_b = 3000\,\mu\text{m/m}$. Valid failure is defined as failure within the free length, while failure close to the clamping section or within the clamping section is deemed invalid.

Results: Testing in Fiber Direction

The testing in fiber direction is performed on a number of $n = 3$ to $n = 5$ specimens at each temperature level, ranging from RT to 250 °C. However, two specimens had defective strain gauges (one at RT and one at 250 °C), which made it impossible to determine the modulus. All specimens showed valid failure modes, as indicated in Fig. 3.4. An attempt to conduct testing at 275 °C was unsuccessful due to specimen slipping. This is interpreted as a consequence of matrix softening, which lead to pressing into the surface and results in slippage of the clamps on the fibers. The determined properties in fiber direction are presented in Fig. 3.5.

Fig. 3.4 Fiber failure modes caused by tensile testing in fiber direction. The nominal length of each specimen shown is 250 mm

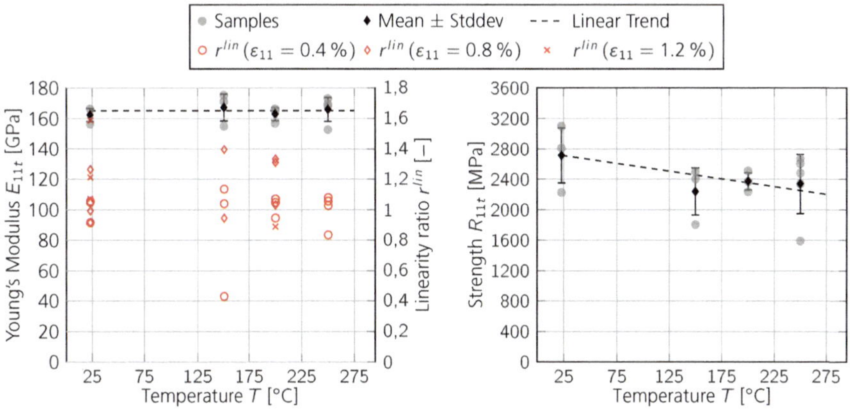

Fig. 3.5 Tensile properties in fiber direction over temperature

The Young's moduli are relatively constant over temperature, ranging from 162 GPa to 168 GPa, which equals a variation of less than 3%. The variation coefficients determined at the individual temperatures are in the range of 2.6% to 5.2%. Due to the coincidence of these magnitudes, the assumption of temperature-independent behavior is reasonable. Furthermore, linear elastic behavior is verified based on the values of the linearity ratio, denoted by r^{lin}, which fall between 0.8 and 1.4. This range is within the acceptable limits of 0 to 2.

The failure modes observed are consistent for all specimen and temperatures, with an explosive-like fracture caused by fiber failure, as shown in Fig. 3.4. The mean strength at RT is determined by about 2713 MPa and decreases with increasing temperature to 2338 MPa at 250 °C. This is approximately 14% lower than the RT value. A linear reduction over temperature is assumed to exist (dashed line), but the trend is not fully reliable due to the limited number of samples available. The variation coefficients of the strength are significantly higher than that of the Young's modulus, being between 13.4% to 16.7% for 23 °C, 150 °C and 250 °C. The relatively low value of about 4.7% found at 200 °C is interpreted as a random occurrence resulting from the low sample number. The relatively high cv_{11t} is a consequence of the inherently cumulative character of uncertainties at peak loads. Detailed inspection of the fractured specimens shows that the high surface pressure applied through the wedge clamping mechanism can cause a first fiber failure and initiate the overall failure. This observation is similar for all tested specimens and at all test temperatures and contributes to the high variability.

Several studies have reported linear tensile strength degradation of CFRP composites with increasing temperature. For instance, Li et al. [5], Zhou et al. [36], Wang et al. [42], and Yu et al. [43] have shown similar trends in their investigations and below T_g. They further report a degressive degradation of tensile strength above the glass transition temperature. This temperature domain is not sufficiently inspected here to confirm such observation for the studied material and in regard to the PES T_g. However, these studies used test equipment that applied the thermal load locally at the specimen center only, keeping the clamped area at a colder temperature. The observation of linear strength degradation below T_g can be attributed to the sensitivity of matrix or interface load transfer capability to temperature. Gibson [44] supposed that fiber-matrix debonding and creep can occur during tensile testing. Several studies, such as those by Gao et al. [45], Subramanian et al. [46], and Reifsnider et al. [47, 48], have interpreted these micromechanical effects using Batdorf et al.'s [49, 50] bundle strength theory to explain the tensile strength degradation due to additional fiber-matrix debonding.

Results: Tensile Testing in Transverse Direction

The transverse tensile testing involves testing of $n = 2$ to $n = 3$ specimens at temperatures up to 275 °C. The partially low number of specimens tested is due to the occurrence of defective strain gauges and preemptive specimen failure. The transverse tensile Young's moduli, E_{22t}, are determined within the measured stress-strain domains which may partially deviate from the standard recommendations due to the preemptive failure. The transverse properties obtained are presented in Fig. 3.6. The

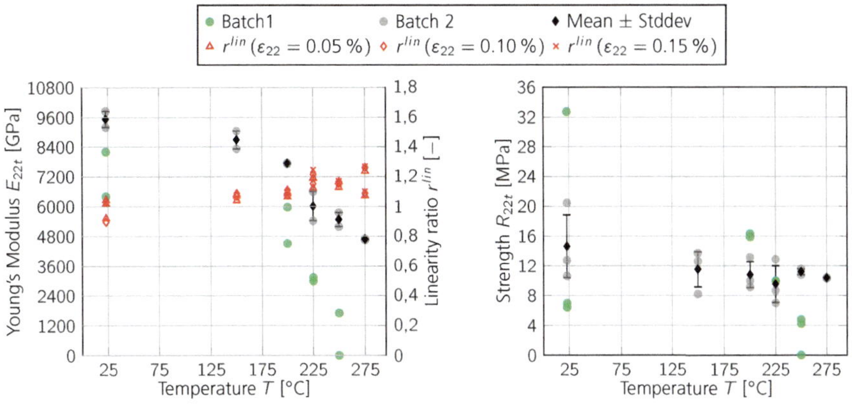

Fig. 3.6 Transverse tension properties of the UD laminate

figure distinguishes two batches of specimens. The first batch utilizes tabbed specimens, while the second batch consists of specimens without tabs. For the purpose of this work, the results of the second batch are used, while the first batch values are used for discussion purposes only within this section.

The Young's modulus of batch 2 is found to be 9524 MPa at RT. As the temperature increases, the modulus decreases with a relatively low slope until it reaches around 200 °C, where the value is approximately 20% lower than that at RT. Beyond 200 °C, the degradation accelerates. Until 275 °C, the modulus magnitude drops to about 50% of its RT value. This change in degradation rate is attributed to the glass transition of the PES toughener material. Noteworthy is, that the tabbed specimens in batch 1 exhibit an even higher drop in elasticity. The significant property scatter at 225 °C is associated with the sensitivity of the glass transition. The r^{lin} confirms linear elastic behavior, but it is important to note that the related strain domain is too small to provide a reliable linearity assessment.

Transverse tensile testing produced partially invalid and premature failure, particularly at temperatures from RT up to 200 °C, where failure occurred consistently within or near the clamping section. However, valid failures are observed at elevated temperatures. This behavior can be attributed to several, probably superposing factors. From a materials standpoint, residual stresses develop between the fiber and matrix during cool-down from the curing temperature, which contributes to the maximum load carrying capability and explains the temperature sensitivity as well as the valid failure at elevated temperature where these stresses are lower. Poor specimen alignment or local surface indentation of the clamps, similar to testing in fiber direction, could also cause local failure initiation, which would explain the failure close to the clamps. Interestingly, one specimen of batch 1 produced significantly higher failure loads at RT compared to the other specimens but still suffered an invalid failure.

The captured strength data shows a similar qualitative trend of strength reduction with temperature compared to the modulus decrease. However, the scatter is high, mostly above 20%, which does not allow for further interpretation. As such, the captured values cannot serve as a reliable database for transverse tensile strength. For future applications, a valid test program must be established to address the limitations encountered during the current testing. A reasonable number of preliminary tests is recommended especially for transverse tensile testing to find a suitable setup. This includes consideration of tailored tabs, careful specimen alignment and clamping to minimize local interferences, and additional characterization of the residual stress state to better understand its contribution to the observed behavior.

3.4.3 Lateral Contraction

Lateral contraction is determined during tensile testing. During compression testing, however, it was not possible to measure due to the small size of the specimen. The ASTM D 3039 standard [30] the strain range for lateral contraction determination equivalent to the strain levels used for determining the Young's modulus. Hence, using strain levels of $\varepsilon_a = 1000\,\mu\text{m/m}$ and $\varepsilon_b = 3000\,\mu\text{m/m}$ the lateral contraction is calculated according to:

$$\nu_{xy} = \frac{\varepsilon_{y,b} - \varepsilon_{y,a}}{\varepsilon_{x,b} - \varepsilon_{x,a}} \tag{3.9}$$

In Eq. 3.9, x and y refer to the load direction applied which is relevant due to the orthotropic character of the UD CFRP lamina. The fiber direction tension tests result in ν_{12} while transverse tensile testing leads to ν_{21} values, as shown in Table 3.2.

3.4.4 Compressive Loading

Test Setup and Procedure

Compression loads can lead deformation normal to the specimen plane causing a stress state which deviates from pure compression. To consider these challenges, various test setups are proposed in standards such as EN 2850 [31] and ASTM D 3410 [51], as well as in the literature. Feih et al. for instance, are using NASA short block

Table 3.2 Poisson ratios determined from tensile testing of UD CE CFRP specimen according to standard ASTM3039

Sign	Unit	RT	200 °C	225 °C	250 °C
ν_{12}	–	0.317	0.372	–	–
ν_{21}	–	0.018	0.023	0.012	0.025–0.04

specimens [52, 53] to determine the compressive properties. However, preliminary studies have found inconsistent applicability of these setups. For the present study, the EN 2850 type B [31] test standard is selected as the most useful and available solution. The test procedure distinguishes between strength determination (type B1) and Young's modulus determination (type B2). Further sample setup specifications are given in Appendix A, Table A.1, and Fig. A.1c–f. In both setups, parallel adapter plates are used to apply compressive loads to the frontal surface of the specimen, as depicted in Fig. 3.7d. Each setup also includes a fixture to align the specimen and to support against deformation out of the laminate plane, as shown in Fig. 3.7a, b, respectively. The type B1 fixture highlighted in Fig. 3.7a is used For strength characterization. This setup uses tabsto ensure good load introduction and to prevent from load edge deformation at high loads. The free length between the tab sections is only 5 mm. In contrast, the Young's modulus determination uses the type B2 fixture as shown in Fig. 3.7b and specimen are not prepared with tabs to allow for a larger free length and sufficient area for strain measurement. Back-to-back strain gauges are used to evaluate the possibility of bending deformation and property falsification. According to the standard, the Young's modulus is determined between 10% and 50% of the load applied, and it is calculated as the average strain of the two back-to-back strain gauges denoted by ε' and ε'':

$$\varepsilon = \frac{\varepsilon' - \varepsilon''}{2} \tag{3.10}$$

Equation 3.11 provides the criterion to check for the presence of bending deformation during compression testing:

$$|\frac{\varepsilon' - \varepsilon''}{\varepsilon' + \varepsilon''}| \leq 5\% \qquad 0.1\sigma_{max} \leq \sigma \leq 0.9\sigma_{max} \tag{3.11}$$

Figure 3.7c displays a thermography image captured during preliminary studies and following the type B1 test procedure. The image was taken after the mounted specimen was removed from the thermal chamber, which was heated to a temperature of 100 °C. As shown in the image, the temperature distribution within the specimen is almost uniform, indicating suitability for testing. To confirm that the desired test temperature is reached, additional thermocouples are attached to the cross section of the specimen.

To determine the compressive strength, the maximum stress that the specimen can withstand is recorded. A valid failure is defined as any failure that occurs within the free length of the specimen, without failure at the specimen ends, according to the standard EN 2850 [31]. The standard ASTM D 3410 [51] provides additional information on possible failure modes, as illustrated in Fig. 3.8.

Results: Testing in Fiber Direction

To determine the properties of the material in the fiber direction, two to four specimen are tested at temperatures up to 250 °C for Young's modulus determination using the type B2 setup. Three to six specimen count is tested at a temperature range of up

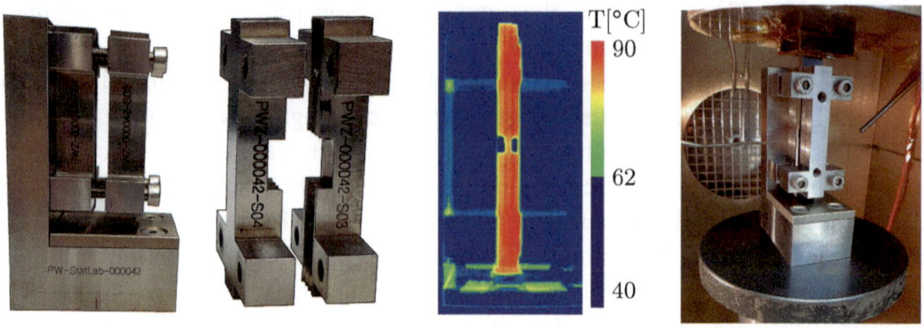

(a) (b) (c) (d)

Fig. 3.7 The compression testing setup includes fixture B1 (**a**), which is used to determine strength properties, and fixture B2 (**b**), which is used to determine elastic properties. Thermography image (**c**) highlights a compression specimen within Fixture B1 shortly after it was removed from the thermal chamber. The fixture is placed on adapter plates for axial loading within the thermal chamber (**d**)

Fig. 3.8 Typical failure modes caused by compression testing [51]. The four left types are assigned as valid while the four right types are assigned as invalid. The figure is extracted from ASTM D3410 [51]

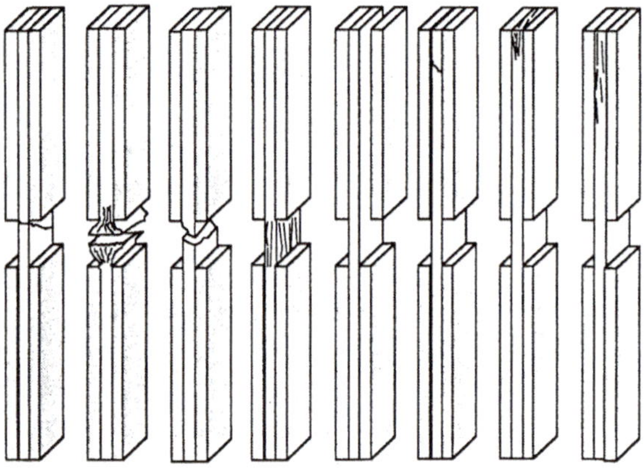

300 °C for determination of the compressive strength using the type B1 setup for capturing the expected strength temperature dependency at temperatures above PES T_g. The resulting properties are shown in Fig. 3.9. Because of the two different test setups, the values of Modulus and Strength are not directly related to each other.

The Young's modulus shows a constant trend up to a temperature of 225 °C within a range of ±5%. It has a magnitude between 140 GPa and 147 GPa with a coefficient of variation below 6%. Considering the coincidence of these magnitudes, the assumption of temperature-independent behavior in this temperature range is reasonable. However, the mean value is roughly 12% lower than the tensile Young's modulus. Above 225 °C, the magnitude of E_{11_c} increases, and the development of bending deformation is visible (see appendix Sect. A.3). This is a result to matrix softening and less fiber support. The elevated Young's modulus is interpreted as a follow-up from such behavior. In any case, the linearity ratio is close to 1, which confirms a linear elastic behavior.

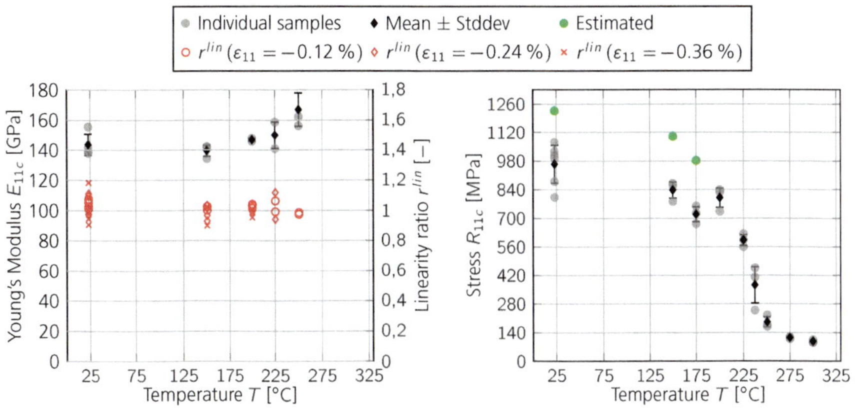

Fig. 3.9 Compression properties in fiber direction

The results of strength testing demonstrate a high sensitivity to tabbing, resulting in mostly invalid failure modes. Different fracture types occurred at different temperature regimes, as shown in Fig. 3.10. At temperatures up to 150 °C, tab delamination led to failure at the specimen ends. At temperatures above this range, a different failure mode occurred. At 225 °C, the failure was initiated within the tab region and propagated towards the free length. At 237 °C, the fracture occurred within the tab region and appeared as a lateral kinking mode, which appeared more ductile than brittle. At all temperatures above 237 °C, valid failures occurred within the free length and the fracture propagated diagonally and seemed to be triggered by the transition from tab to free length.

Figure 3.11 provides a comparison of the fracture types that developed at 200 °C and above at the microscopic scale. At 200 °C and 225 °C, the fracture developed as a brittle fiber failure mode, partially connected to local delaminations. At 250 °C, the failure appeared differently in a more ductile kinking mode. A delamination separated the fibers, forming kink-bands at different locations without breaking them. This observation underlines the loss of fiber support as a consequence of low matrix softening. Feih et al. [52, 53] reported a similar failure mode switch to kinking. Gibson et al. [44, 54] also observed kinking and shear bands, which explain the degradation of compressive strength. Although these studies are made for biaxial fabric material, these observations are consistent with the present findings and underline the sensitivity resulting to matrix softening.

The measured strength values, as presented in Fig. 3.9, exhibit a temperature-dependent relationship similar to the transverse DMA curve. The most significant property drop occurs within the range of PES T_g, between 200 °C and 250 °C. The highest variation coefficients are observed at RT, which is sensitive to tab debonding, and within the aforementioned glass transition domain. The variation coefficients for all other temperatures are relatively stable at around 5–6%.

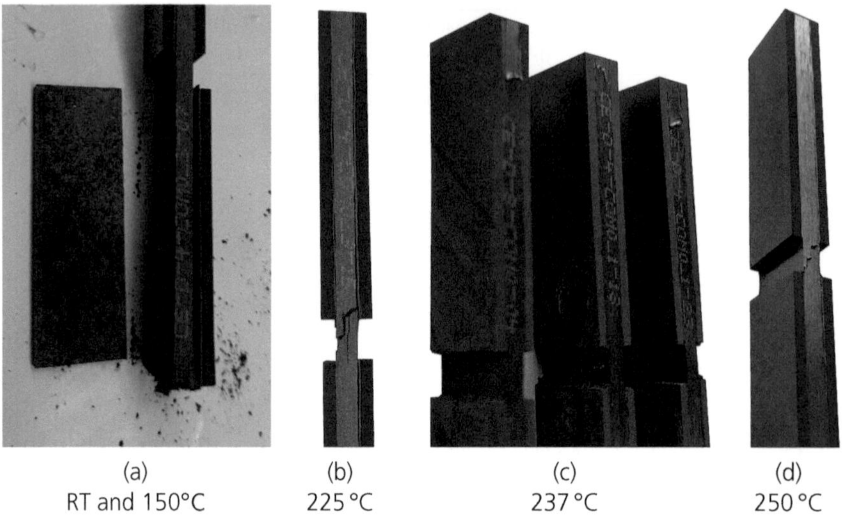

Fig. 3.10 UD specimen failure modes observed from compression loading in fiber direction at different test temperatures. (**a**) Specimen end failure after tab delamination, (**b**) central failure, partially triggered by tab delamination, (**c**) lateral kinking (**d**) valid failure

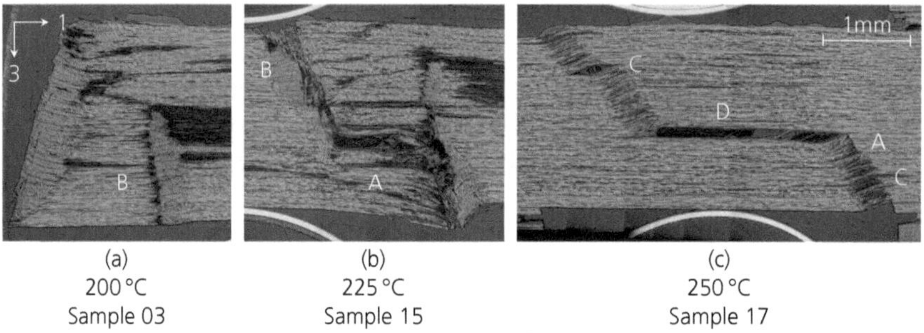

Fig. 3.11 Microscopy images of composite fractures under longitudinal compression loads at different temperatures. Letters describe different mechanisms. A Brittle kinking in normal direction; B Brittle kinking in lateral direction; C Soft kinking; D Delamination

However, due to the predominance of invalid failure modes in the tests, some considerations are necessary when applying R_{11_c} in modelling. The measured values below 200 °C follow the overall trend but represent lower magnitudes. Due to the preemptive specimen end failure, the measured strength in this temperature range underestimates the actual material strength. A representative compression strength margin is found within the T800-S carbon fiber datasheet [25], which is the carbon fiber used within the tCFRP material, providing a compression strength of $R_{11_c} =$ 1228 MPa for an epoxy composite. This value is used to estimate the offset value between the datasheet and actual values, assuming an ideal case of fiber failure within the free length. Three data points are estimated adding this offset to the measurement data. These points are presented as green circles in Fig. 3.9 located at temperatures up to 200 °C. It should be noted that the sensitivity of the matrix system to failure is neglected in this estimation, which may cause deviating behavior between epoxy and tCE. Using these estimated points (green), the overall strength degradation follows a more continuous degradation that is similar to a typical s-shape. The studies conducted by Gibson et al. [44, 54] obtained similar strength degradation for glass fabric composites, with a significant decrease observed around the T_g. These findings support the trends observed in this study.

Results: Transverse Compression Testing

Transverse compression testing is conducted using two to five specimens at temperatures up to 300 °C for strength and up to 250 °C for the Young's modulus. Again, the strength test is done at even higher temperatures to get a better view on its degradatation behavior. The resulting properties are presented in Fig. 3.12. In contrast to compression testing in the fiber direction, transverse compression testing produced valid failure behavior in all cases.

To improve the accuracy of the Young's modulus determination and avoid errors due to initial stress development, the stress range for modulus evaluation was adjusted to 30–70% in this study. The linearity ratio (r^{lin}) exhibits values greater than 1, particularly at elevated temperatures, which can be attributed to the initial stress buildup. After this initial increase in load, the r^{lin} remains relatively constant, indicating a linear response. The measured nonlinearities are within acceptable limits and refer mostly to the full strain range up to failure. Additionally, all tests yielded valid and reproducible behavior, with a bending ratio within standard specifications. At RT, the Young's modulus is approximately 9700 MPa with a variation coefficient of approximately 3%. As the temperature rises, there is an almost linear reduction in the Young's modulus, resulting in a value below 50% of the RT value at 250 °C.

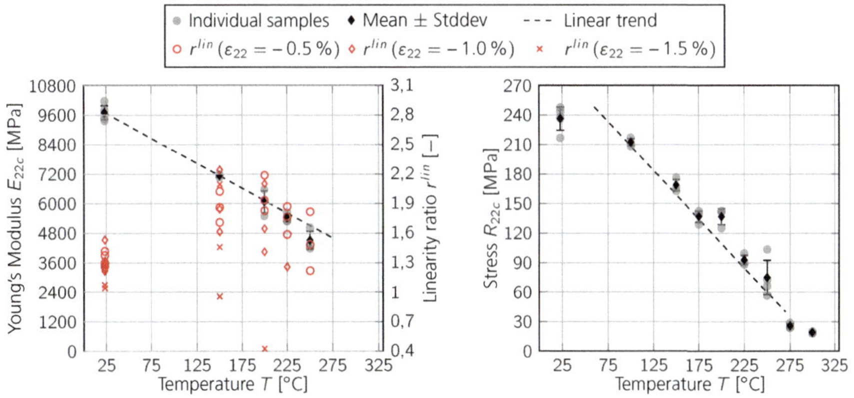

Fig. 3.12 Transverse compression properties of the UD laminate

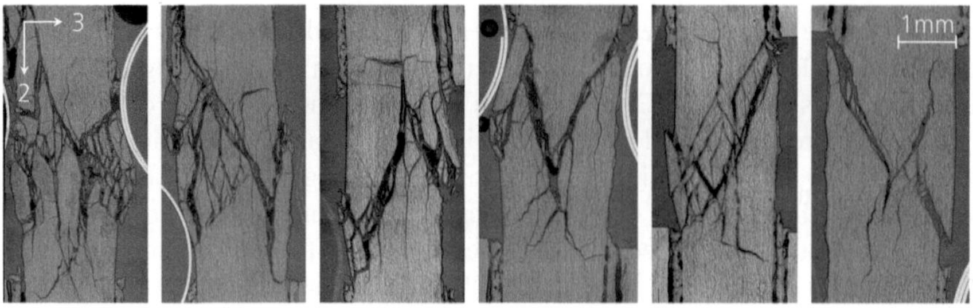

(a) 23 °C Sample 55 | (b) 100 °C Sample 69 | (c) 150 °C Sample 57 | (d) 200 °C Sample 60 | (e) 225 °C Sample 50 | (f) 250 °C Sample 63

Fig. 3.13 Fracture surfaces observed after transverse compressive testing

Figure 3.13 presents microscopy images comparing fractures at different temperatures. These images show that all transverse compression strength tests resulted in valid failure modes within the free length. At RT, the fracture surface appears brittle with one major inclined failure plane and many small, brick-like fragments near the surface. Similar failure planes are visible at all temperatures, but the size and amount of smaller fragments decrease with increasing temperature. At 250 °C, only the failure plane and some minor crack tips are visible. This behavior visualizes the matrix softening. As shown in Fig. 3.12, the strength values continuously decrease with increasing temperature, and the trend can be interpreted as a s-shape with more continuous degradation compared to the R_{11_c} behavior. The highest variation coefficient is obtained at 250 °C, again caused by the PES glass transition. The variation coefficients for all other temperatures are around 5%. Above 250 °C, the strength magnitude approaches a plateau, and the variation returns to the typical magnitude. Although more data would be required to confirm this trend, this observation is interpreted as a follow-up to a complemented PES glass transition.

3.4.5 *In-plane Shear Loading*

The IPS test setup used in this study follows the standard ASTM D 3518 [32]. It employs a simple tensile test setup using $\pm 45°$ laminate specimens, which generates a uniform stress state over the thickness and along the specimen [55, 56]. Although another test setup called the Iosipescu test (or V-notched specimen test) is also available for determining shear properties, as for instance applied by Gentz et al. [57] at RT and 316 °C, it requires precise specimen geometry [26]. In contrast, the IPS test is simpler leading to less sensitivities of the experimental setup and should therefore lead to less uncertainty within the results. Hence, it is selected for the present study. Notably, tabs are not used. Shear stress τ_{12} and shear strain γ_{12} are calculated from axial and transverse strain measurements, ε_x and ε_y, respectively, according to the standard ASTM D 3518 [32]. For a detailed overview of the test setup, see Table A.1.

$$\tau_{12} = \frac{F}{2A} \quad ; \quad \gamma_{12} = \varepsilon_x - \varepsilon_y \tag{3.12}$$

To determine the shear modulus G_{12}, Eq. 3.1 is applied using τ_{12} and γ_{12} at shear strains of $\gamma_{12,a} = 1500\,\mu\text{m/m}$ and $\gamma_{12,b} = 5500\,\mu\text{m/m}$.

There are different procedures to determine the shear strength, including the typical maximum stress criterion that determines $R_{12,max}$. According to ASTM D 3518 [32], the maximum stress must be obtained at or below 5% shear strain. As a result of shear deformation, the specimen undergoes lateral contraction, resulting in a plateau within the stress-strain relation. This plateau represents the maximum shear stress level, and the corresponding strength is referred to as $R_{12,ult}$. An alternative method for strength determination is the offset shear strength criterion, where the strength is defined by the intersection of the stress-strain curve and a straight line obtained by the shear modulus which is shifted by a predefined strain offset value. ASTM D 3518 [32] recommends using a strain offset value of $0.2\% = 2000\,\mu\text{m/m}$. A similar procedure is also used for linearity assessment, as explained in Sect. 3.3.3, where a linearity ratio of zero equals the intersection point. The corresponding strength value is designated as $R_{12,0.2}$. Table A.1 provides a detailed overview of the test setup.

Results

Three to four specimens are tested in a temperature range between RT and 250 °C. The resulting properties are shown in Fig. 3.14. The fracture behavior observed at ultimate failure is similar across all temperatures and is indicated in Fig. 3.15. However, at elevated temperatures, strain gauges failed at lower strain magnitudes compared to RT (see Appendix Sect. A.4). This did not affect the modulus determination, but for

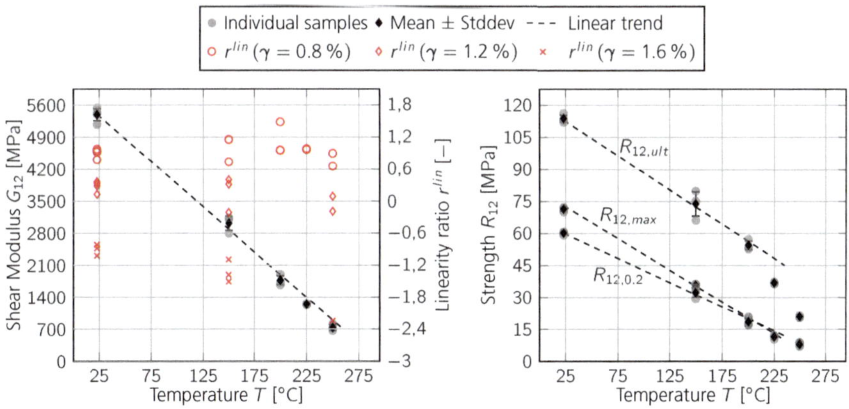

Fig. 3.14 Characterization results for in-plane shear testing: Elastic behavior (left) and Strength behavior (right)

Fig. 3.15 Shear failure at ultimate loads produced at 250 °C (top) and 23 °C (bottom)

strength determination, a second-degree polynomial extrapolation has to be applied in some cases to estimate the stress-strain plateau. To facilitate better understanding, the polynomials have been added to the raw data charts in the appendix (Sect. A.4), and the related strength values have been designated accordingly (see Table A.11). It has to be emphasized that these results are therefore not fully reliable but serve here for studying the trends over temperature.

Stress-strain curves with typical progression for shear loads are obtained at RT, and the shear modulus is derived to be approximately 5400 MPa within the linear regime. At 150 °C, the typical behavior is still observed, but the curve proceeds significantly lower. This results in a reduction of the shear modulus by 56% compared to the RT shear modulus value which is a severely lower value compared to the other moduli. However, the modulus is still determined within the linear domain, as indicated by the linearity ratio ($r^{lin} > 0$). The shear modulus degradation is almost linear over temperature, reaching only 13.6% at 250 °C. In [58], the shear modulus shows a similar continuous degradation over temperature. The shear modulus of all tests at high temperatures is still determined within the designated linear regime. However, the stress-strain relations clearly follow a nonlinear path from the first loading onwards which is also visible by the trend of the linearity ratio at high strains. These observations raises the question if the strain offset margins are suitable. However, this would require further testing, as indicated earlier (conf. Sect. 3.3.3). Gsell et al. [11] investigated the sensitivity between yield strength and temperature, explaining these trends. The best convergence between multiple tests and in reference to the stress-strain relation is found at RT and 225 °C. They describe variation coefficients below 2.5%. All other curves deviate somewhat, with the highest variation found at 250 °C.

The shear strength magnitudes $R_{12,max}$ and $R_{12,0.2}$ degrade over temperature similar to the shear modulus. The mean values relative to their RT value, $\tilde{\mu}^{RT}$, are always slightly below the shear modulus counterpart. A clear drop around the PES glass transition is not visible and can only be found at the ultimate strength values R_{12}^{ult}. With the exception of the measurement at 200 °C, the variation coefficients of R_{12}^{max} are relatively low, with magnitudes below 3%. Similar low values are found for the ultimate strength, while the offset strength, $R_{12,0.2}$, clearly shows a certain sensitivity that results from the modulus determination (offset line definition) including strain measurement and higher variation coefficients.

3.4.6 Interlaminar Shear Loading

The interlaminar shear strength (ILSS) is a measure of the resistance of a laminate against shear failure between adjacent layers. Although it is not a property that is required for the structural analysis with the intended material theory of thin-walled shells, it is of general interest within the scope of the present program because it is temperature sensitive and depends on the state of the matrix and interface. The test is performed on a UD laminate, in accordance with the standard EN 2563 [33], using a three-point bending setup. The bending is applied along the fiber direction, resulting in a non-uniform strain ε_{11} across the thickness of the specimen. Specimen alignment is relatively easy to achieve, and the test does not require strain measurement, making it simple to perform. The apparent shear stress τ_{13} is calculated from the applied force F and the specimen cross-sectional area A as indicated in Eq. 3.13. The shear strength is finally obtained by determining the maximum stress.

$$\tau_{13} = \frac{3 \cdot F}{4 \cdot A} \qquad \rightarrow \qquad R_{13} = max\,(\tau_{13}) \tag{3.13}$$

A valid failure in the ILSS test is defined as a brittle shear failure mode, which may occur due to delamination in one or multiple planes. However, if the failure mode is a bending failure, i.e. fiber failure at the specimen surface side, or the deformation behavior is plastic, the Eq. 3.13 becomes invalid at the moment of failure [33]. Therefore, such failure modes are considered invalid and are verified using both the load-displacement curve and visual inspection. A plastic failure leads to a gradual decline in the load-displacement relation, which is smoother compared to the abrupt spikes that result in case of a delamination or bending failure.

Results

Tests are carried out on three to seven specimens per temperature within the temperature range of RT to 300 °C. An overview of the results is presented in Fig. 3.16, while the raw data can be found in the appendix in Sect. A.5.

At RT, the predominant failure mode is delamination in multiple planes. The load-displacement curves exhibit multiple small spikes until a more severe drop in load is observed, which indicates the ILSS failure and defines R_{13}. The microscopy image in Fig. 3.17a shows four distinct delaminations. A similar behavior is observed for temperatures up to 150 °C, as evident from the load-displacement behavior and fractures depicted in Fig. 3.17b. At 175 °C, the load-displacement relation becomes more nonlinear but still consists of spikes representing delamination onset. However, at 200 °C, spikes are absent but a distinct load drop is still observed. At higher temperatures, the behavior becomes more nonlinear. The macroscopic deformation is plastic and irreversible, and microscopy images, such as Fig. 3.17c, show numerous delaminations. The onset of plasticity is observed between 200 °C and 225 °C, as seen in Fig. 3.16. In this temperature range, the R_{13} transitions from an almost linear behavior at lower temperatures to a higher slope, thereby exhibiting a similar property drop as observed for transverse properties. A certain S-shape towards

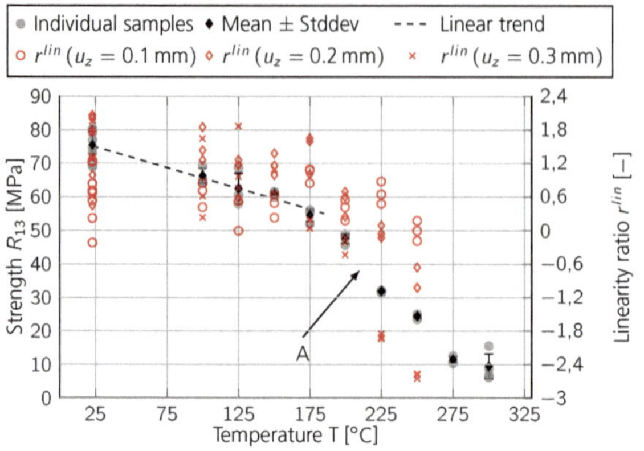

Fig. 3.16 ILSS determined at several temperature levels according to standard EN 2563 [33]; A: Onset of plastic failure mode (visually inspected)

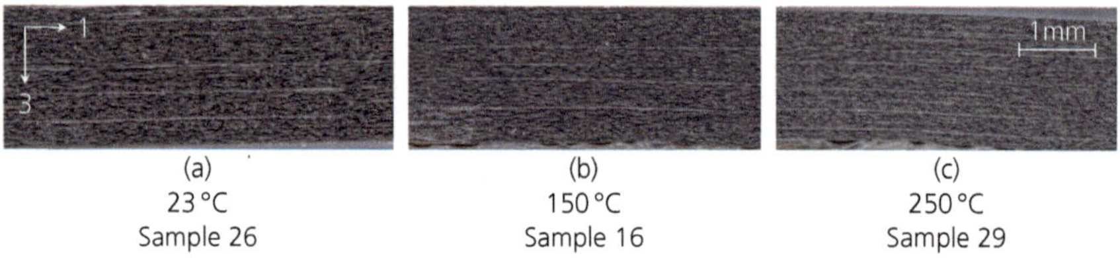

Fig. 3.17 UD specimen failure caused by ILSS testing. Delaminations are visible as horizontal, bright gray lines, representing gaps between adjacent CFRP layers

a constant property level beyond 225 °C is also suggested. The linearity ratio also displays a changing behavior, starting around 200 °C. It is determined through the load-displacement relation here and has a relatively high variation and can only be regarded as indicator of a trend to support the findings. The variation coefficients are relatively small, with 5.5% at RT and below 3.5% for the majority of test temperatures. Above 250 °C, the variation increases significantly, which appears to be a result of the overall changing behavior at PES T_g and related uncertainties.

3.4.7 *Thermal Expansion Behavior*

Changing temperature causes a change in volume for unrestrained specimen. At a uniform temperature change, i.e., extension or contraction in each direction, the dimensional change can be described by the thermal strain vector $\underline{\varepsilon}$th, which consists of the thermal strain components in individual directions. For calculating the thermal strain, the thermal expansion characteristics of the material must be known which are, similar to the mechanical properties, obtained in orthotropic form for the UD lamina.

The material properties that describe the dimensional change due to temperature are expressed by the CTE's denoted by α_{ij}. Therefore, the functional relation to calculate the thermal strain can be written as:

$$\varepsilon_{ij}^{th} = \alpha_{ij} \Delta T \tag{3.14}$$

The measurement of CTE's is performed using a TMA (thermomechanical analysis) setup. In this setup, a rod is used to apply a constant but small force, typically $0.05N$, to the sample to ensure continuous contact. The specimen is then heated, and its resulting dimensional change is measured through the displacement of the rod, Δu. Aligning the specimen in different position leads to measure the different CTE components. Due to the small force, stress development is negligible. Simultaneous temperature measurement allows for the determination of the CTE using Eq. 3.15 [59]:

$$\alpha_{ij} = \frac{\Delta u_{ij}}{u_{ij}^{0} \cdot \Delta T} \tag{3.15}$$

In the present study, measurements of CTE's are conducted for each spatial orientation of the UD composite, namely fiber direction, transverse direction, and thickness direction. Figure 3.18 presents the resulting rod displacement over temperature in gray. Assuming a constant thermal expansion behavior over temperature, linear regression is used to derive the CTE's, indicated by the black dashed lines. The measurements in transverse and thickness directions apply two slopes to distinguish the CTE's at temperatures below PES T_g and above. An overview of the resulting values is provided in Table 3.3. An analytical calculation of the properties applying homogenization is conducted to compare with the measured values.[4] In reference to the study of Bowles and Tompkins [61], the equations of Shapery [62] and Chamis [6] are used for homogenizing the α_{11} and α_{22} properties, respectively.

The measurements in the fiber direction exhibit highly varying deflections. The curves shown in Fig. 3.18 (left) show three different samples. All samples show noisy signals which are attributed to the devices working close to their operating limits. Further, spikes are observed in different shape between the individual specimen which are sensitive to the evaluation and are likely a result of surface tolerances and rod settlement. The α_{11} values resulting to these measurements are relatively low, but still higher than both estimations. The absence of a changing slope around T_g indicates the fiber dominance.

In contrast, measurements in the transverse and thickness directions show a much smoother dimensional change with a higher measurement domain and less sensitivity to surface tolerances resulting in a relatively low variation. The measurements below T_g exhibit a good correlation with the estimation. Above T_g, the magnitudes of both

[4] The magnitudes of the transverse fiber CTE are obtained from Kulkarni and Ochoa [60], and are applied within the range of $\alpha_{22f} \in \{5.6 \cdot 10^{-6} \frac{1}{\mathrm{K}}; 11.5 \cdot 10^{-6} \frac{1}{\mathrm{K}}\}$. The longitudinal fiber and resin properties are listed in Table 3.1 and are used accordingly.

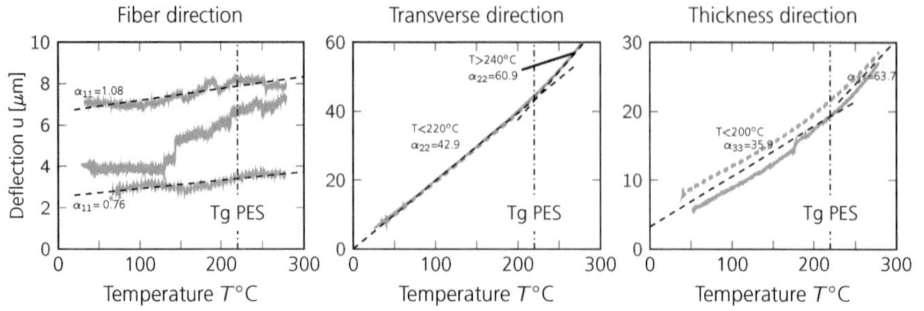

Fig. 3.18 TMA measurement results. Gray lines represent the measured dimensional change. Linear regression shown in black dashed lines represent the resulting CTE's

Table 3.3 Comparison of measured and predicted CTE. The prediction assumes equal values in transverse and thickness direction ($\alpha_T = \alpha_{22} = \alpha_{33}$)

Sign		α_{11} $[10^{-6}\frac{1}{K}]$	α_{22} $[10^{-6}\frac{1}{K}]$	α_{33} $[10^{-6}\frac{1}{K}]$
μ	T< T_g	0.92	42.9	35.9
$\mu_{estimated}$		0.23	37.5 to 42.0	
s		0.16	0.4	1.8
μ	T> T_g	0.92	60.9	63.7
s		0.16	0.7	4.0

α_{22} and α_{33} increase by about 50%, which is due to the changing resin properties and the matrix dominance within these directions.

3.5 Evaluation of the Temperature Dependent Behavior

The previous section provides a comprehensive overview of the material characterization results. Effective material properties are determined according to typical standards at various discrete temperature levels, and the material behavior is discussed within the context of individual tests and experiences found in the literature. The properties are characterized for the UD ply and serve as a database for the material modelling, which formulates a functional relation of the properties over temperature and is explained in Chap. 4. The present section compares and discusses the captured properties in terms of their sensitivity to temperature. The following topics are addressed:

- The sensitivity of residual stress to the characterization results
- Elastic behavior over temperature and verification of linear elasticity
- Strength behavior over temperature
- Material uncertainties.

Residual Stresses

The curing process of composite materials results in residual stress, a phenomenon that arises from the manufacturing stage. Due to the heterogeneity of composite materials, different amounts of strain are induced within its constituents due to chemical shrinkage during curing and thermal stress during cool-down after curing. This strain difference between the constituents is finally translated into stress, which is known as residual stress. According to Cuntze [63, 64], two types of residual stress are relevant for composite failure. Residual stress at the meso-scale, i.e. within the UD ply, occurs when the individually oriented plies are constrained and stressed by the overall Multidirectional (MD) laminate deformation. Residual stress on the micro-scale arises from the different behavior of fiber and matrix. The presence of residual stress and thermal stresses is a critical factor in the thermomechanical condition of the material and is sensitive to the material behavior. The actual magnitude of stress that is present within a material depends on the residual stress from curing, the thermal state and resulting thermal stress and the mechanical loads applied. This highlights the link between residual stress and thermal state as well as the sensitivity to the mechanical behavior.

An influential effect observed throughout characterization is the debonding of tabs. It can be attributed to the development of residual and thermal stress. In these cases, the tab and specimen have different laminate stackings and thermal expansion behavior. They are bonded using the tCE material and the associated curing process. During the cooling process from the curing temperature to room temperature, the thin tCE adhesive film is significantly stressed. Simultaneously, a poor surface connection and low-quality adhesive layer can result in tab debonding, which is observed in many specimen. This highlights the importance of considering residual and thermal stress in characterization programs, particularly under when varying thermal conditions are investigated. The characterization program considered the materials in their UD form and tested until limit conditions, which is why such stress can become severely sensitive to the characterization result. For example, Naik [65] demonstrated that residual stress can exceed the predicted transverse tensile strength. To overcome the tab debonding issue, the use similar laminates is recommended and the consideration of beveled tabs may be beneficial to reduce stress peaks. However, such setups should be verified before planning extensive programs.

The influence of residual stress occurring at the micro-scale is believed to be highly sensitive to the characterization results. Studies by Hobbybrunken et al. [40] and Brauner [66] concluded on significant tensile stress states to develop within the fiber-matrix interface and matrix. Kulkarni [60] further emphasizes the sensitivity of transverse failure initiated by such stress. In addition, fiber-matrix disbonding has been reported by Lüders [12]. Assuming that the fiber governs its thermal expansion to the matrix, a simplified estimation yields tensile matrix stress ranging from 12.4 MPa to 18.4 MPa, which is up to 40% of the neat CE strength.[5] These issues are inherent to FRP composite materials. However, investigating these phenomena

[5] Material properties used are listed in Table 3.1. The transverse fiber properties used are given by Kulkarni and Ochoa [60], see footnote 3.4.7. Estimation leads to a CTE mismatch of $30 \cdot 10^{-6} \frac{1}{K}$

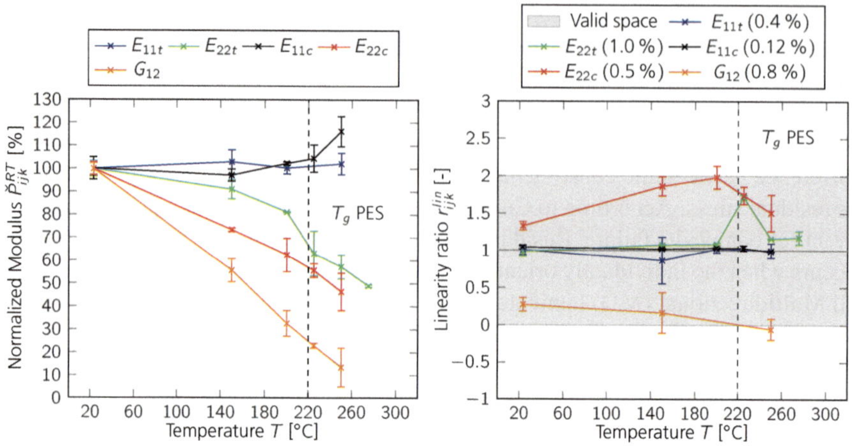

Fig. 3.19 Evaluation of the experimentally determined elastic behavior in dependence of temperature for the UD lamina. The moduli are normalized to their RT quantity. The linearity ratios refer to different strain levels as indicated within the legend. The gray domain refers to the 0.2% strain offset margin

experimentally is not easy [63]. Nevertheless, these reflections help to understand the problems encountered during transverse tensile characterization and explain a source of property scatter in general.

The material characterization is sensitive to residual stress for particular areas. However, this aspect is not investigated in detail in this work due to its focus. Nevertheless, this conclusion highlights that it is crucial to account for residual stress development at the ply-level (meso-scale) when studying laminate behavior and, ultimately, the behavior of structures under thermomechanical conditions. Section 5.1.4 provides a more detailed explanation of this point.

Elastic Behavior

Figure 3.19 presents a comparison of the individual elastic properties behavior over temperature. The left chart shows the moduli normalized to their values at RT. The right chart displays the linearity ratios observed with respect to the reference strain domain indicated in the legend. The gray shaded region indicates the range in which the assumption of linearity is valid, as explained on Sect. 3.3.3. In both charts, the dashed black line highlights the T_g of the PES toughener.

The Young's moduli for tensile and compressive loads in the fiber direction remain relatively constant over temperature. The tensile modulus has a magnitude of $E_{11_t} = 165$ GPa, while the compressive modulus has a magnitude of $E_{11_c} = 145$ GPa, which is about 12.1% lower than the tensile modulus. The variation of mean values over temperature is mostly within the range of property uncertainty, except for the single compressive modulus value at 250 °C. This confirms that assuming a

and $20 \cdot 10^{-6} \frac{1}{\text{K}}$ between fiber and matrix and results in thermal strain within the matrix between 0.32% and 0.48%.

constant property behavior over temperature is a valid approximation. The linearity ratio of both moduli remains close to 1, confirming linear elasticity. These two observations are relevant input for modelling.

The transverse Young's moduli for both tensile and compressive loading are similar at RT, with values of approximately 9.5 GPa and 9.7 GPa, respectively. The shear modulus has a magnitude of about 5.4 GPa at RT. However, all of these moduli show a clear temperature dependency, with individual trends that diverge from each other. The E_{22_t} and E_{22_c} deviate until 200 °C but converge at higher temperatures. The shear modulus gradually decreases with temperature and at magnitudes below. The curves of the compression modulus and shear modulus show almost linear degradation over temperature, while the tensile modulus exhibits a drop before reaching the T_g of the PES toughener. This deviation could be explained by different mechanisms at the microscale, which ultimately cause the temperature dependency of the resulting property at the macroscale. However, this phenomenon requires further investigation and verification.

The determined linearity ratios confirm a reasonable linear-elastic behavior for all of these moduli. However, for the transverse compression modulus and the shear modulus, the trends approach the top and bottom border, respectively, with increasing temperature. This behavior is particularly expected for the shear modulus, as such behavior is known to develop nonlinear and plastic behavior that intensifies at elevated temperatures [11]. It is important to note that these linearity ratios refer only to a specific domain of strain, and higher strain domains could lead to more nonlinearity and plasticity. This effect could also become increasingly severe with temperature as for instance indicated in [44, 54]. Therefore, further investigation capturing the full strain domain is required to verify these expectations. A relevant future advancement would also be the use of a more adequate criterion for assessing the onset of nonlinearity and plasticity, as proposed by Henry et al. [38]. As proposed earlier, this could include a proper selection of the offset strain magnitude or other, more appropriate test procedures.

Strength Behavior

Figure 3.20 presents the comparison of the individual strength property behavior over temperature. The left chart shows the UD ply strength components normalized to their magnitude at RT. The right chart shows the ratio of strength to modulus, η_{ijk} for the individual components, as obtained by Eq. 3.16. This relation is a simplified (one-dimensional, neglection of other components) approximate of the failure strain and the combined temperature dependency. In both charts, the dashed lines indicate the T_g of the PES toughener.

$$\eta_{ijk} = \frac{R_{ijk}}{E_{ijk}} \sim \varepsilon_{ijk}^{ult} \tag{3.16}$$

The strength properties in the fiber direction exhibit significantly different magnitudes. The tensile strength is approximately $R_{11_t} = 2713$ MPa, while the compression strength, R_{11_c}, is around 1228 MPa. It should be noted that the compression strength

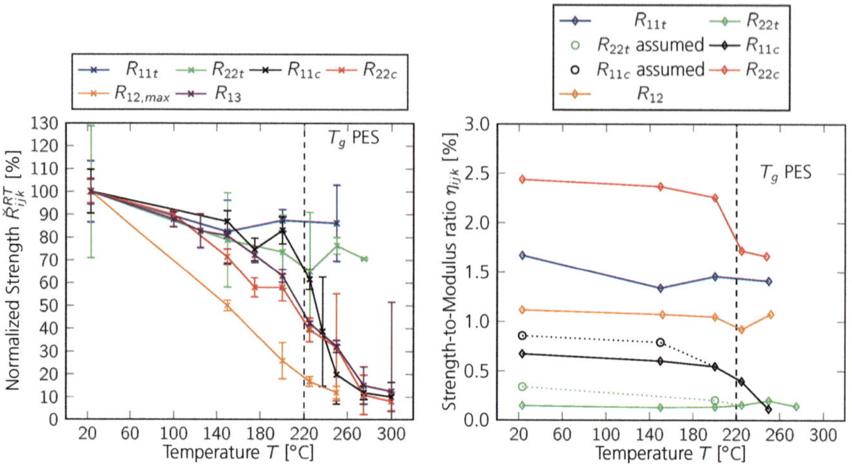

Fig. 3.20 Evaluation of the UD laminate strength behavior over temperature determined by static material characterization: Values normalized to their RT quantity (left) and effective strength ratios (right)

is a predicted value due to invalid specimen end failure (see Sect. 3.4.4). The property degradation is relatively smooth and in both cases similar until about 200 °C. However, at higher temperatures, the tensile strength continues its linear degradation trend, while the compression strength clearly drops. This indicates the sensitivity resulting from matrix softening and related loss of fiber support. For compression, the failure mode changes in this temperature domain, with a visually softer fracture that includes delamination and kinking at temperatures above the PES T_g. These observations reveal a much more severe temperature dependency for the compression strength, which aligns with the findings of Gibson et al. [44, 54]. The trends of the η values are similar to the strength component, since the modulus values do not exhibit significant deviations.

The transverse strength components and the ILSS exhibit similar trends over temperature as the compression strength in fiber direction, suggesting a similar temperature dependency. However, the transverse tensile testing produced invalid failure modes at low temperatures, making its trend over temperature unreliable. The degradation of the R_{22_c} and R_{13} is smoother compared to that of the R_{11_c}, which can be attributed to the different failure phenomenology in relation to the matrix softening. The behavior of the η_{22_c} is relatively constant until 200 °C. The distinct drop at higher temperature suggests that matrix softening is more severe at elevated compression loads that are outside the domain of modulus determination. The shear strength degradation follows a similar trend as its corresponding modulus degradation, resulting in an almost horizontal η_{12} trend.

The observed trends and failure characteristics provide valuable insights into the behavior of the material under varying temperature conditions. It is important to note that some of the tests did not yield valid failure modes, and assumptions had to be made in these cases. Despite these limitations, the results are consistent with the expected degradation behavior and are sufficient input for the material modelling. However, to gain a more comprehensive understanding of the failure mechanisms and trends, further investigations with a higher number of test temperatures and improved statistical representativity are necessary.

Material Uncertainties

In structural design, material uncertainties are accounted for by using material allowables. In this work, material allowable models that depend on temperature are developed, which therefore must consider the material uncertainties as well, in accordance with Working Hypothesis 1 (see Sect. 2.3). To capture these uncertainties, the variation coefficients are evaluated for each property and test temperature of the characterization program. However, it should be noted that the number of test repetitions does not lead to a reliable statistically basis and a higher number of tests would be required. Hence, the measured variation coefficients provide only a preliminary approximation. Therefore, it is not possible to derive a Probability Density Function (PDF) for the material property scatter, and it is assumed to be normally distributed. This assumption is used for any property scatter in this work. An overview of statistical distributions of material properties is provided in [67]. The Military Handbook MIL-HDBK-17-1F [68] provides methods and guidelines for testing datasets to determine their distribution, and is a recommended reference for deriving PDF's.

There is limited information available in the literature regarding the scatter behavior of composite materials over temperature. Sriramula and Chryssanthopoulos [67] provide a comprehensive review of material uncertainty quantification and an overview of uncertainties in various FRP materials, but no information is provided on the temperature dependency of these uncertainties. To provide an understanding of how individual properties may scatter, a comparative overview of selected studies is presented in Table 3.4. These studies refer to different types of epoxy CFRP materials tested at RT conditions, as no information is available for a material solution similar to the one tested in this work. Jeong and Shenoi [69] used 35 specimens per property, while Lin et al. [70, 71] used 17 specimens per property. Degenhardt et al. [72] did not report on sample numbers but included three individual characterization programs of the same material. Camanho et al. [73] used the same material as Degenhardt and conducted five repetitions per property. To compare the uncertainties of the present characterization to the literature values, an average variation coefficient is calculated.

$$\bar{cv}_P = \frac{1}{\sum n(T)} \sum cv(T) \cdot n(T) \tag{3.17}$$

The use of an average variation coefficient assumes a consistent scatter behavior over temperature and smears out deviations that could occur at individual temperatures. It must further be noted that this comparison has limitations as the studies refer to different material solutions and test conditions. Figure 3.21 displays the comparison of variation coefficients of the elastic properties (left) and the strength

Table 3.4 Comparison of variation coefficients for different materials given in the literature and referring to RT conditions of the present study. Mat1: T300/DMS2224; Mat2: IM7/8552; Mat3: Toho Q-1115

$cv\,(P)$	[69]	[72]	[72]	[72]	[73]	[70, 71]	Present study	
%	Mat1	Mat2				Mat3	$\bar{cv}^P\,(CE)$	sel.
E_{11_t}	4.6	1.2	3.1	1.4	1.4	3.6	4.0	4.0
E_{22_t}	4.0	2.4	3.9	2.9	1.0	4.3	3.9	4.0
E_{11_c}	–	1.8	1.7	2.4	1.4	–	4.3	4.0
E_{22_c}	–	6.8	4.9	4.1	1.0	–	4.3	4.0
G_{12}	3.1	2.3	13.6	1.1	2.5	5.5	4.4	4.0
ν_{12}	6.0	5.6	13.5	–	6.2	5.9	–	–
R_{11_t}	14.3	3.4	11.9	3.6	5.8	2.2	12.8	10.0
R_{22_t}	11.1	18.6	18.1	26.5	8.5	4.3	29.0	12.0
R_{11_c}	13.5	4.9	9.0	7.2	12.1	2.2	8.5	8.0
R_{22_c}	7.4	6.6	5.1	6.0	10.2	4.3	6.9	7.0
R_{12}	4.1	4.6	17.5	0.8	0.7	5.9	6.2	6.0
R_{13}	2.4	–	–	–	–	5.9	4.1	4.0

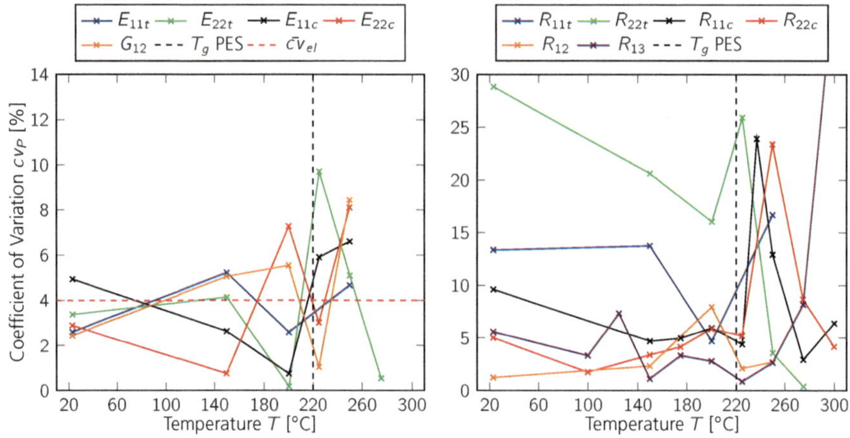

Fig. 3.21 Uncertainties of the present characterization program

properties (right) over temperature. The T_g of the PES toughener is indicated by a black dashed line. Table 3.4 compares the average variation coefficients to the values reported in the literature, and also includes a column indicating the magnitude of scatter selected for use in material modelling.

The variations in all elastic properties, range from 3% to about 5% until approximately 200 °C. Up to this temperature, the trends remain mostly constant or decrease. An increase in scatter is noticed within the PES T_g domain. The values found in the literature are generally consistent with each other and in a similar domain compared to the tCFRP values, particularly in the case of Jeong and Shenoi [69] and Lin et al. [70, 71], who used a large number of specimens. Therefore, a $\bar{cv} = 4.0\%$ is selected to represent the material uncertainty for all elastic properties and over the entire temperature range and is used within material modelling.

Generally, the strength properties display larger deviations between individual properties compared to the elastic properties. The trends of the strength properties decrease until around 200 °C, after which they increase significantly, indicating high sensitivity to temperature. This is expected because of the sensitivity to failure initiation to specific matrix or interface properties. Moreover, matrix material behavior introduces further sources of scatter at high loads, such as plasticity, and is dependent on temperature. The transverse tensile strength shows particularly high variation coefficients which may additionally be caused by the reported test difficulties. However, similarly high magnitudes are also reported in the literature and underline the sensitivity of the above described source of scatter on the transverse tension strength property. Table 3.4 shows the variation coefficients selected for use in material modelling, which is done based on the actual measurement and the literature values. The highest magnitudes are in tensile conditions, while the others are still above the variation coefficients of elastic properties.

3.6 Summary and Conclusion

To investigate the structural behavior under distributed thermal conditions, it is essential to understand the material properties in dependence on temperature. However, experimental characterization of these properties poses a challenge due to the difficulties involved in testing with temperature and the significant quantity of testing required at different temperatures. This chapter presents a material characterization program that addresses these challenges and captures the required input for deriving temperature-dependent property models. This contribution aligns with working hypothesis 1 and the overall research objective. The resulting property models are discussed in Chap. 4.

The characterization program uses common test standards and performs testing at a reasonable number of test temperatures and the total testing effort is relatively low, considering that the properties are known over the full temperature range.

The studies presented in this chapter focus on a particular UD CFRP material solution that consists of a toughened cyanate ester matrix (CE). CE is a high-temperature resistant material that can withstand temperatures up to 400 °C. To increase the material's fracture toughness, PES additives are used. Such material solutions have not been investigated in the extent of this program before. However, these additives significantly affect the material properties over temperature, and their sensitivity varies depending on the load direction in the UD laminate. The characterization program captures these temperature dependencies for both the elastic and strength properties. The main aspects include:

- The Young's moduli in the fiber direction are constant until around 250 °C. Higher temperatures are expected to cause modulus degradation.
- The Young's moduli in the transverse direction and for shear loads are significantly sensitive to temperature.
- Linear-elastic behavior is verified within the temperature and strain range investigated in this study. However, temperature-sensitive effects such as the development of plasticity may occur outside this range, particularly at higher strains and at high temperature. Therefore, further investigations are needed to investigate these domains.
- The material behaves relatively brittle and exhibits relatively low fracture strains. It was not possible to determine the ultimate strains due to measurement limitations.
- The tensile strength in the fiber direction exhibits a relatively small degradation up to 250 °C, indicating typical fiber failure. Similar degradation is observed for compressive loads up to 200 °C, followed by a significant drop due to matrix softening and reduced fiber support, leading to a failure mode switch.
- The transverse tensile strength is observed with invalid failure modes at low temperatures, with failure occurring at the load introduction area, while at elevated temperatures, valid failure is observed.
- The transverse compression strength exhibits valid failure modes within the tested temperature range until 300 °C. The sensitivity to temperature is similar to that of compression modulus and the property drops around the PES T_g
- The shear strength and shear modulus show a similar relatively linear dependency on temperature.
- ILSS exhibits a similar transition as the transverse compression properties. A failure mode transition towards plastic failure occurs at around 200 °C
- The scatter of properties over temperature is found to show a significant increase around T_g for all properties. It should be noted, however, that this is only a preliminary investigation that is based on a low statistical basis.

References

1. N. Platzer, Progress in polymer engineering. Eng. Chem. **61**(5), 10–30 (1969)
2. C.A. Mahieux, K.L. Reifsnider, Property modeling across transition temperatures in polymers: a robust stiffness-temperature model. Polymer **8**(42), 3281–3291 (2001). https://doi.org/10.1023/A:1011230620286
3. M.F. Ashby, *Engineering Materials 2: An Introduction to Microstructures and Processing*, 4th edn. International Series on Materials Science and Technology. (Butterworth-Heinemann, Oxford, 2012)
4. C. Sauder, J. Lamon, R. Pailler, Thermomechanical properties of carbon fibres at high temperatures (up to 2000 °C). Composit. Sci. Technol. **62**(4), 499–504 (2002). ISSN: 0266-3538. https://doi.org/10.1016/S0266-3538(01)00140-3
5. C. Li, G. Xian, Mechanical property evolution and life prediction of carbon fiber and pultruded carbon fiber reinforced polymer plate exposed to elevated temperatures. Polym. Compos. **41**(12), 5143–5155 (2020). https://doi.org/10.1002/pc.25782

6. C.C. Chamis, Simplified Composite Micromechanics Equations for hygral, Thermal and Mechanical Properties. NASA Technical Memorandum 83320. Houston, Texas (1983). https://ntrs.nasa.gov/citations/19830011546
7. C.C. Chamis, Mechanics of Composite materials: Past, Present and Future. NASA Technical Memorandum 100793. Lewis Research Center, Cleveland, Ohio (1984). https://ntrs.nasa.gov/citations/19880008360
8. H. Altenbach, J. Altenbach, R.B. Rikards, *Einf ührung in die Mechanik der Laminat- und Sandwichtragwerke: Modellierung und Berechnung von Balken und Platten aus Verbundwerkstoffen* (Verlag f ür Grundstofflndustrie, Dt, 1996). 3-342-00681-1
9. M. Liebisch, T. Wille, Sucohs Project. ECCOMAS Thematic Workshop: Computational Multi-Physics, Multi-Disciplinary and Multi-Data - CM3 (2021)
10. C. Mittelstedt, W. Becker, Strukturmechanik ebener Laminate. 1. Auflage. Darmstadt (2016). ISBN: 978-3-935868-99-0
11. C. G'sell, D. Jacques, J.P. Favre, Plastic behaviour under simple shear of thermosetting resins for fibre composite matrices. J. Mater. Sci. **25**(4), 2004–2010 (1990). https://doi.org/10.1007/BF01045756
12. C. L üders, Mehrskalige Betrachtung des Erm üdungsverhaltens thermisch zyklierter Faserkunststoffverbunde. Dissertation. TU Braunschweig (2020)
13. P. Haupt, On the mathematical modelling of material behavior in continuum mechanics. Acta Mech. **100**(3–4), 129–154 (1993). https://doi.org/10.1007/BF01174786
14. M. Liebisch, Structural design concepts map for thermal and fibre loaded composite structures. Sucohs project, Deliverable 3.1 (2020). https://www.sucohs-project.eu/publications-presentations
15. A. Baltopoulos et al., Multifunctional properties of multi-wall carbon nanotubes/cyanate-ester nanocomposites and CFRPs, in *SPIE Proceedings*, ed. by J. Leng, A.K. Asundi, and W. Ecke (2009). 74932G. https://doi.org/10.1117/12.845657
16. A. Inamdar et al., Thermoplastic-toughened high-temperature cyanate esters and their application in advanced composites. Ind. Eng. Chem. Res. **57**(13), 4479–4504 (2018). https://doi.org/10.1021/acs.iecr.7b05202
17. Lonza, Primaset PT-30. A Multifunctional Cyanate Ester Delivering High Tg and Inherent Flame Retardancy: Technical Datasheet (2015)
18. K.N. Shivakumar, H. Chen, G. Holloway, Effect of thermal fatigue on tensile and flexural properties of carbon/cyanate ester pultruded composite. J. Reinf. Plast. Compos. **28**(6), 675–689 (2009). https://doi.org/10.1177/0731684407086614
19. A.J. Kinloch, A.C. Taylor, The toughening of cyanate-ester polymers Part I Physical modification using particles, fibres and woven-mats. J. Mater. Sci. **37**, 433–460 (2002)
20. A.J. Kinloch, A.C. Taylor, The toughening of cyanate-ester polymers: Part II Chemical modification. J. Mater. Sci. **38**, 65–79 (2003). https://doi.org/10.1023/A:1021109731672
21. L. Amirova, F. Schadt, C. Brauner, M. Grob, Properties and structure of Cyanate Ester/Polyethersulfone composites. Athens (2019). https://www.sucohs-project.eu/publications-presentations
22. F. Schadt, L. Amirova, C. Brauner, M. Grob, Thermoplastic particle toughening of Carbon fiber/ Cyanate Ester composites. Athens (2019). https://www.sucohs-project.eu/sites/sucohs/files/media/2019-10/SuCoHS_EASN2019_Presentation_Schadt_20190905.pdf
23. Amsler und Frey AG, Polyethersulfon: Technical Datasheet (2022). www.amsler-frey.ch
24. L. Amirova et al., Properties and structure of high temperature resistant cyanate ester/polyethersulfone blends using solvent-free toughening approach. Polymer Bulletin (2020). ISSN: 0170-0839.https://doi.org/10.1007/s00289-020-03493-w
25. Toray Composite Materials America, Inc., T800S intermediate modulus carbon fiber: Technical-Data-Sheet (2021). www.toraycma.com. Accessed from 24 Apr 2021
26. L.A. Carlsson, R.B. Pipes, *Experimental Characterization of Advanced Composite Materials*, 2nd edn. (Technomic Publ, Lancaster, Pa, 1997). ISBN: 1-56676-433-5. http://www.loc.gov/catdir/enhancements/fy0744/96061023-d.html

27. M. Liebisch, *Temperaturabhängige Materialeigenschaften f ür die Struktursimulation thermomechanisch gekoppelter Anwendungen: Hochleistungstechnologie Thermoanalyse - Neue Messmethoden und wissenschaftliche Anwendungen der Luft- und Raumfahrt* (An-wendung von thermoanalytischen Kenngrößen in der Luft- und Raumfahrt, Tagung, 2018)
28. M. Liebisch, Increased exploitation of CFRP composites for aerospace structures under thermomechanical loading. ZwickRoell Forum for High-Temperature Testing (2019)
29. M. Liebisch et al., Thermomechanical Testing of CFRP Structures under Varying Thermal Conditions. SAMPE EUROPE Conference and Exhibition 2021, Baden, Switzerland (2021)
30. ASTM International, Standard Test Method for Tensile Properties of Polymer Matrix Composite Materials (2007)
31. DIN-Normenausschuss Luft- und Raumfahrt, Unidirektionale Laminate aus Kohlenstoffasern und Reaktionsharz - Druckversuche parallel zur Faserrichtung (2018)
32. ASTM International, Standard Test Method for In-Plane Shear Response of PMC-Materials by Tensile Test of a +-45°-Laminate (2001)
33. Normenstelle Luftfahrt im DIN Deutsches Institut f ür Normung e.V. DIN EN 2563: Kohlenstoff-faserverstärkte Kunststoffe - Unidirektionale Laminate - Bestimmung der scheinbaren inter-laminaren Scherfestigkeit (1997)
34. H.B.M. HBM, Serie KFU, Elektrische Dehnungsmessstreifen, Datenblatt, B04924 (2018). https://www.hbm.com/de/
35. L. Ost, Screening und Sensitivitätsanalyse von Einflussfaktoren auf das Strukturverhalten thermomechanisch gekoppelt beanspruchter CFKStrukturen: DLR-Interner Bericht. DLR-IB-FA-BS-2020-162. Masterarbeit. Cottbus-Senftenberg: Brandenburgische Technische Universität (2019). https://elib.dlr.de/139559/
36. F. Zhou et al., Effect of temperature on material properties of carbon fiber reinforced polymer (CFRP) tendons: experiments and model assessment. Materials (Basel, Switzerland) **12**(7)(2019). ISSN: 1996-1944. https://doi.org/10.3390/ma12071025
37. Normenstelle Luftfahrt im DIN Deutsches Institut f ür Normung e.V. Nichtmetallische Werkstoffe - Vorzugspr üftemperaturen (1992)
38. C. Henry et al., Evaluation of an Alternate Method for Determining Yield Strength Offset Values for Selective Laser Sintered Polymeric Materials (2019). https://doi.org/10.33599/nasampe/s.19.1583, https://www.nasampe.org/store/ListProducts.aspx?catid=&ftr=10.33599/nasampe/s.19.1583
39. T. Hobbiebrunken et al., Thermomechanical analysis of micromechanical formation of residual stresses and initial matrix failure in CFRP. JSME Int. J. Ser. A Solid Mech. Mater. Eng. **47**(3), 349–356 (2004). https://doi.org/10.1299/jsmea.47.349
40. T. Hobbybrunken et al., Microscopic yielding of CF/epoxy composites and the effect on the formation of thermal residual stresses. Compos. Sci. Technol. **65**(10), 1626–1635 (2005). https://doi.org/10.1016/j.compscitech.2005.02.003
41. R. Hein, Vorhersage und In-Situ Bewertung fertigungsbedingter Deformationen und Eigenspan-nungen von Kompositen. Dissertation. TU Braunschweig (2019). https://elib.dlr.de/128886/
42. Y.C. Wang, P. Wong, V. Kodur, An experimental study of the mechanical properties of fibre reinforced polymer (FRP) and steel reinforcing bars at elevated temperatures. Compos. Struct. **80**(1), 131–140 (2007). ISSN: 02638223. https://doi.org/10.1016/j.compstruct.2006.04.069
43. B. Yu, V. Kodur, Effect of temperature on strength and stiffness properties of near-surface mounted FRP reinforcement. Compos. Part B: Eng. **58**, 510–517 (2014). ISSN: 13598368. https://doi.org/10.1016/j.compositesb.2013.10.055
44. A.G. Gibson et al., High temperature and fire behaviour of continuous glass fibre/polypropylene laminates. Compos. Part A: Appl. Sci. Manuf. **41**(9), 1219–1231 (2010). ISSN: 1359835X. https://doi.org/10.1016/j.compositesa.2010.05.004
45. Z. Gao, K.L. Reifsnider, Micromechanics of tensile strength in composite systems. Compos. Mater.: Fatigue Fract. **4**, 453–470 (1993). https://doi.org/10.1520/STP24745S
46. S. Subramanian, K.L. Reifsnider, W.W. Stinchcomb, Tensile strength of unidirectional composites: the role of efficiency and strength of fiber-matrix interface. J. Compos. Tech. Res. **17**(4), 289–300 (1995). https://doi.org/10.1520/CTR10449J

47. K.L. Reifsnider, S. Case, Mechanics of temperature-driven long-term environmental degradation of polymer-based composite systems. Proc. ASME Mater. Div. **69**(1) (1995)
48. K.L. Reifsnider et al., Mechanics of Long-Term Behavior of High Temperature Polymer Composites (1998). https://apps.dtic.mil/sti/citations/ADA356865
49. S.B. Batdorf, Tensile strength of unidirectionally reinforced composites - I. J. Reinforced Plast. Compos. **1**(2), 153–164 (1982). ISSN: 0731-6844. https://doi.org/10.1177/073168448200100206
50. S.B. Batdorf, R. Ghaffarian, Tensile strength of unidirectionally reinforced composites -II. J. Reinforced Plast. Compos. **1**(2), 165–176 (1982). ISSN: 0731-6844. https://doi.org/10.1177/073168448200100207
51. D30 Committee, Test Method for Compressive Properties of Polymer Matrix Composite Materials with Unsupported Gage Section by Shear Loading. West Conshohocken, PA. https://doi.org/10.1520/D3410_D3410M-03R08
52. S. Feih et al., Modelling the tension and compression strengths of polymer laminates in fire. Compos. Sci. Technol. **67**(3–4), 51–564 (2007). ISSN: 02663538. https://doi.org/10.1016/j.compscitech.2006.07.038
53. S. Feih et al., Modelling the compression strength of polymer laminates in fire. Compos. Part A: Appl. Sci. Manuf. **38**(11), 2354–2365 (2007). ISSN: 1359835X. https://doi.org/10.1016/j.compositesa.2007.04.013
54. A.G. Gibson et al., Modeling composite high temperature behavior and fire response under load. J. Compos. Mater. **46**(16), 2005–2022 (2011). ISSN: 0021-9983. https://doi.org/10.1177/0021998311429383
55. P.H. Petit. A simplifled method of determining the inplane shear stress-strain response of unidirectional composites. *Composite Materials: Testing and Design*, ed. by S. Yurenka. (ASTM International, 1969), pp. 83–93. ISBN: 978-0-8031-0017-6. https://doi.org/10.1520/STP49808S
56. B.W. Rosen, A simple procedure for experimental determination of the longitudinal shear modulus of unidirectional composites. J. Compos. Mater. **6**(3), 552–554 (1972)
57. M. Gentz et al., In-plane shear testing of medium and high modulus woven graphite fiber reinforced/polyimide composites. Composit. Sci. Technol. **64**(2), 203–220 (2004). ISSN: 02663538. https://doi.org/10.1016/S0266-3538(03)00260-4
58. A.G. Gibson, Laminate theory analysis of composites under load in fire. J. Composit. Mater. **40**(7), 639–658 (2005). ISSN: 0021-9983. https://doi.org/10.1177/0021998305055543.
59. International Standard, Plastics - Thermomechanical analysis (TMA): Determination of coefficient of linear thermal expansion and glass transition temperature (2009)
60. R. Kulkarni , O. Ochoa, Transverse and longitudinal CTE measurements of carbon fibers and their impact on interfacial residual stresses in composites. J. Composit. Mater. **40**(8), 733–754 (2005). ISSN: 0021-9983. https://doi.org/10.1177/0021998305055545
61. D.E. Bowles, S.S. Tompkins, Prediction of coefficients of thermal expansion for unidirectional composites. J. Composit. Mater. **23**, 370–388 (1989). ISSN: 0021-9983. https://doi.org/10.1177/002199838902300405
62. R.A. Schapery, Thermal expansion coefficients of composite materials based on energy principles. J. Composit. Mater. **2**(3), 380–404 (1968). ISSN: 0021-9983. https://doi.org/10.1177/002199836800200308
63. R.G. Cuntze, The predictive capability of failure mode concept-based strength criteria for multi-directional laminates-part B. Composit. Sci. Technol. **64**(3–4), 487–516 (2004). ISSN: 02663538. https://doi.org/10.1016/S0266-3538(03)00225-2
64. R.G. Cuntze, A. Freund, The predictive capability of failure mode concept-based strength criteria for multidirectional laminates. Composit. Sci. Technol. **64**(3–4), 343–377 (2004). ISSN: 02663538. https://doi.org/10.1016/S0266-3538(03)00218-5
65. N.K. Naik, R. Anbarasu, G.V. Kumar, Effect of curing stresses on the behavior of fibre reinforced plastic composites under biaxial loading. J. Mater. Sci. **26**(72), 5753–5761 (1991). https://doi.org/10.1007/BF01130111

66. C. Brauner, Analysis of process-induced distortions and residual stresses of composite structures. Dissertation. Universität Bremen. ISBN: 978-3-83-25-3528-5
67. S. Sriramula, M.K. Chryssanthopoulos, Quantiflcation of uncertainty modelling in stochastic analysis of FRP composites. Compos. A Appl. Sci. Manuf. **40**(11), 1673–1684 (2009). https://doi.org/10.1016/j.compositesa.2009.08.020
68. Composite Materials Handbook: Volume1. Polymer matrix composites guidelines for characterization of structural materials. Vol. MIL-HDBK-17-1F. (2002). https://appsdtic.mil/sti/citations/ADA426516
69. H.K. Jeong, R.A. Shenoi, Probabilistic strength analysis of rectangular FRP plates using Monte Carlo simulation. Comput. Struct. **76**(1–3), 219–235 (2000). ISSN: 00457949. https://doi.org/10.1016/S0045-7949(99)00171-6
70. S. Lin, Reliability predictions of laminated composite plates with random system parameters. Probab. Eng. Mech. **15**(4), 327–338. (2000) ISSN: 02668920. https://doi.org/10.1016/S0266-8920(99)00034-X
71. S. Lin, T. Kam, Probabilistic failure analysis of transversely loaded laminated composite plates using first-order second moment method. J. Eng. Mech. **126**(8), 812–820 (2000). https://doi.org/10.1061/(ASCE)07339399(2000)126:8(812)
72. R. Degenhardt et al., Investigations on imperfection sensitivity and deduction of improved knock-down factors for unstiffened CFRP cylindrical shells. Composit. Struct. **92**(8), 939–1946 (2010). ISSN: 02638223. https://doi.org/10.1016/j.compstruct.2009.12.014
73. P.P. Camanho, P. Maimí, C.G. Dávila, Prediction of size effects in notched laminates using continuum damage mechanics. Composit. Sci. Technol. **67**(13), 2715–2727 (2007). ISSN: 02663538. https://doi.org/10.1016/j.compscitech.2007.02.005

Chapter 4
Modelling of Temperature Dependent Composite Properties

This chapter presents the development of material models, which are functions of each material property in relation to temperature. These models are based on the characterization results provided in the previous chapter and are therefore created for the tCFRP material solution. The material models serve as the foundation for structural analysis with detailed, spatially distributed temperatures, as they enable the assignment of material properties at any temperature within the considered range.

4.1 Scope of Material Modelling

It is important to note that material models have a specific range of validity and may not be applicable beyond that range. From a practice-oriented standpoint, certain model aspects are more important to consider while others can be neglected. Therefore, this section provides an overview of the scope of material model development, including limitations and practical considerations.

The experimental data presented in Chap. 3 provides material properties that capture the behavior of the UD lamina. These properties homogenize the heterogeneous behavior of the fibers and matrix into a meso-scale formulation that captures the effective orthotropic behavior of the composite. For simplicity, a linear-elastic approach is used, neglecting nonlinear stress-strain relations. The models adopt these simplifications governed by the characterization. The modeling must consider the temperature dependence of each individual material property since the orthotropy exhibits different sensitivities for fiber- and matrix-dominated properties. The characterization data contains all properties that are relevant to consider a plane state of stress. Thereby, the models satisfy the conditions for applying the CLT and to analyze MD laminates. This is essential for analyzing thermal strains at the macro

M. Liebisch, *Design Load Allowables for Composite Plates Exposed to Thermomechanical Loads*, Mechanics and Adaptronics,
https://doi.org/10.1007/978-3-031-71352-1_4

scale, which arise from temperature changes as a result of mutual constraints within MD laminates and it is further a crucial detail for predicting deformation and failure under thermomechanical conditions.

The avoidance of detrimental effects is an important requirement in aircraft design, as they can have a significant impact on material failure behavior (conf. Fig. 2.3). Reversible material behavior is a key consideration of the models, as it verifies the avoidance of such effects. The experimental characterization program aims to correlate such detrimental effects with nonlinear stress-strain relationships, which indicate effects such as matrix yielding and plasticity, interfacial debonding, and individual fiber cracks. Linear elasticity confirmed the absence of these effects. Neglecting matrix plasticity is a common assumption at low temperatures, but it becomes relevant at elevated temperatures, as indicated by several studies [1, 2]. Although the experimental assessment confirmed linear elasticity, the strain domain investigated is insufficiently represented in these studies. While reversibility is assumed here, it is therefore recommended to conduct more detailed investigations of these detrimental effects, especially at elevated temperatures, to verify this assumption.

Safety margins are important in the design process for ensuring conservatism. Well-established engineering definitions, such as typical material allowables, take into account inherent property scatter (see Sect. 2.1). For the overall objective, it is therefore essential to develop a similar formulation for property models over temperature. Given the limited number of samples, the presumption of normally distributed properties is made to address the issue of statistical representativity in the measurements and the variation coeffients are presumed to be constant over temperature, conf. Sect. 3.5 and Table 3.4. It is also important to address model uncertainty since the derived models may not be exact. This is needed to ensure a conservative property model formulation for a robust structural design. Therefore, it is an essential task to quantify and to consider model errors in the material modeling procedure. This task refers to Working Hypothesis 1.

4.2 Procedure of Material Modelling

This section presents an overview to the behavior of polymers in dependence of temperature. This behavior forms the foundation for modelling the properties of polymers and FRP. Various approaches for modelling the behavior of polymers are discussed and provide the mathematical framework for material modeling.

4.2.1 Polymer Behavior in Dependence of Temperature

The temperature dependency of polymers results to physical principles that lead to a changing microstructure with changing temperature. As described by Ashby [3], Maxhieux [4] or Bai et al. [5] the condition of a polymer can be divided into different

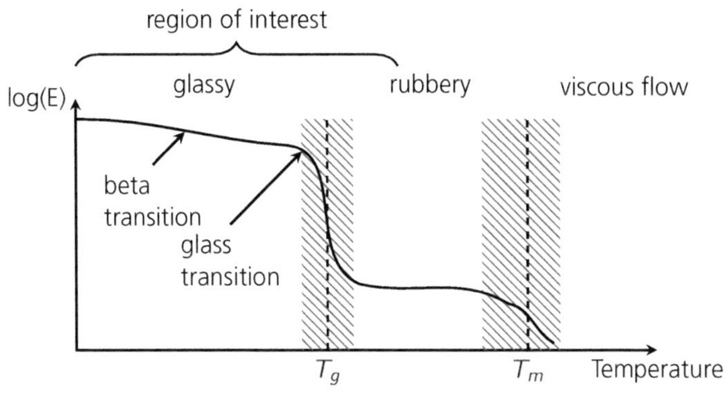

Fig. 4.1 Generic visualization of the polymer softening process over temperature. The figure is freely adapted from [4]

states over temperature, typically ranging from a glassy state towards a decomposed state. Each material state is attributed to a specific microstructure and related features causing the macroscopic properties. A schematic overview to an exemplaric property-temperature relation is presented in Fig. 4.1. The figure shows different states and highlights relevant transition points which refer to a certain temperature range.

Within the glassy state, the material has a solid consistency with high elasticity. However, exceeding the glass transition temperature (T_g) leads to significant mechanical property degradation caused by breakage of secondary bonds. The value of T_g represents the related transition temperature and marks the upper limit of the glassy state temperature range. However, the polymer microstructure consists of different types of bonds [4, 5]. Secondary bonds are relatively weak connections (hydrogen bonds, dipole interaction, Van der Waals interaction) with lower dissociation energy, while primary bonds (e.g. covalent intra-molecular bonds, cross-links of thermosets) have much stronger connections. Hence, multiple transition temperatures with different degradation magnitude can exist depending on the microstructure of the polymer.[1] In addition to the glass transition, the example in Fig. 4.1 highlights an additional property relaxation that is observed within the glassy temperature regime. It is designated here as the beta-transition.[2] This beta transition step is caused by lower energy bond dissociation and minor polymer chain rearrangement and has a smaller degradation magnitude which is smoothly spread over a wider temperature domain. A similar behavior is observed for the investigated CE (see Sect. 3.2) and it is therefore relevant for the material modelling. However, these transitions do not change the microstructure in its entity and therefore do not allow to address a certain additional polymer state.

The rubbery regime develops at temperature that follows the glassy state. Materials in rubbery state have significantly lower elasticity compared to a glassy state, as the

[1] For more specific details the author refers to the cited literature.

[2] Similar observations are commonly made for plastics, and the designation used here is selected in reference to that. However, it should not be taken to indicate any specific chemical reaction.

bonds that transfer the majority of the load are broken. As such restrictions given by the bonds are not active, the polymer chains have more freedom to move. The rubbery temperature regime is, therefore, characterized by interactions between the molecular chains and their entanglement. Some secondary bonds are left, and thus a low amount of elasticity is kept. To provide a complete picture of the temperature-dependency of polymer properties, it should be noted that further increases in temperature beyond the melting point T_m can lead to dissociation of primary bonds, a significant drop in elasticity, and ultimately to viscous flow and decomposition. However, this is beyond the scope of this thesis and is not further discussed.

The material models of this work consider the temperature domains of the glassy state, the glass transition domain and the initial part of the rubbery state. The description above explains the phenomenological behavior of a polymer and is relevant for modelling the individual properties of the composite. Finally, the material models are specifically derived for the selected UD tCFRP composite material. As a consequence of fiber alignment and constant fiber properties over temperature, the polymer behavior causes sensitivities which vary for the individual orthotropic properties of the composite. These sensitivities has to be captured by the models.

4.2.2 Overview to Selected Model Approaches

Bai et al.'s model [6, 7] represents the material as a homogeneous entity comprising different states, each of which is determined by a unique term that defines its conversion α as a function of temperature T. This term is formulated using the Arrhenius approach (as shown in Eq. 4.1) with A, E_A, and n representing the pre-exponential factor, activation energy, and reaction order, respectively and assuming a constant temperature rate β to disregard time dependency.

$$P \propto \frac{d\alpha}{dT} = \frac{A}{\beta} exp\left(-\frac{E_A}{RT}\right) \cdot (1-\alpha)^n \tag{4.1}$$

The kinetic model parameters can be determined through techniques such as thermogravimetric analysis (TGA) measurements [8] or DMA curves [6]. Functions of the material properties in dependence of temperature are obtained by superposing the individual state terms. The model's thermal response and its application to strength properties are discussed in [5, 9], respectively, with good agreement reported between the model values and experimental data. However, the model's parameter dependencies and associated uncertainties are complex, making it unsuitable for the present thesis.[3]

[3] Nevertheless, the approach may prove useful for specific applications, such as investigating the structural behavior in fire scenarios where the material exhibits Arrhenius-like decomposition behavior and time-dependent load conditions. However, a careful study of the correlation between mechanical properties and model parameters is necessary to ensure the accuracy and reliability of the results.

Platzer [10] has described how the strength and rigidity of plastics depend primarily on the interaction between the polymeric chains. This confirms the descriptions of the polymeric softening process around Fig. 4.1. Mahieux and Reifsnider [4, 11–17] developed an approach that is, in similarity to Bai's approach, using the principle of superposition. However, their mathematical formulation accounts for the presence of different inter- and intramolecular bond types and their sensitivity to state transitions.[4] In their theoretical model, a state change is built on the relation between bond failure probability and temperature. Temperature is considered as an energy level, and it correlates with the amount of acting molecular motion [18]. With increasing temperature, an increasing number of interactions between molecular chains or chain segments occurs, leading to an increased statistical chance of bond rupture with temperature. The approach differentiates bond types regarding their temperature domain (or dissociation energy level) that leads to a state change. The superposition considers, that multiple bonds act at the same time and may dissociate (transition) step-by-step over the range of temperature. The macroscopic mechanical properties of a polymer are then linked to the condition of the bonds. Finall, the mathematical model expressions allows to differentiate according to existing bond types, their strength, and their dissociation processes and the formulation is useful for describing the relation between temperature and mechanical properties.

As explained around Fig. 4.1, the glassy state is characterized by multiple polymer bond types, stronger ones and weaker ones. Each of them can stretch or break, resulting in a quasi-static behavior of the polymer [11]. The mathematical approach used to model the property development over a state transition, as explained above, is based on viscoelastic spring-dashpot models [4], where a property relaxes over time. The equation used for this approach is:

$$P\left(t\right)=\sum_{r=1}^{n}P^{r}exp\left(\frac{-t}{\tau^{r}}\right) \tag{4.2}$$

where P represents the general property of interest, t is the time variable and τ^r is the relaxation time constant. In this context, properties are arranged in a parallel series, where each member is associated with a specific bond and represents a portion of the overall property.

As indicated earlier, the relationship between property relaxation and temperature can be analogized to the relationship between polymer chain reptation and energy level. Therefore, an Arrhenius-like model, as proposed by Bai, seems logical, and the similarity between Eq. 4.2 and the Arrhenius equation (Eq. 4.1) becomes evident. The approach uses the exponential nature of the Boltzmann distribution. However, to account for the progressive interaction between molecular chain motion and bond rupture, Mahieux and Reifsnider propose a Weibull distribution as a more appropriate form to model the probability of bond breakage with increasing temperature and the related property relaxation. Thus, Eq. 4.2 is modified by introducing the Weibull

[4] For the sake of simplicity, the following explanation uses the general term of *bond* to refer to both bond and interaction types, without differentiation or discussion between them.

shape parameter m, which is an exponential factor. Introducing the time-temperature analogy by substitution of time t with temperature T leads to the general version of Mahieux's and Reifsnider's model approach:

$$P(T) = \sum_{r=1,s=r+1}^{n} H^{rs} exp\left(-\left(\frac{T}{T^{rs}}\right)^{m^{rs}}\right) \quad \text{with} \quad H^{rs} = P^r - P^s \tag{4.3}$$

The term used within the sum of Eq. 4.3 represents a single transition process. The superposition of multiple transitions, as provided by sum formulation, leads to the general model equation. In the present discussion, each bond type is associated with an individual transition step and refers to an individual term. Such transition occurs between two material states, r and s and is based on its dissociation temperature and the related bond mechanism. The quantity of property degradation H^{rs}, the model parameter m^{rs} and the transition temperature T^{rs} are introduced to scale the mechanical property for a transition rs, where P^r and P^s are the relaxed properties before (unbroken bonds) and after the transition (broken bonds), respectively. Figure 4.2 illustrates the sensitivities of individual parameters for a single transition step (n = 1).

The sensitivity of a transition step to the macroscopic property is captured by the parameter H^{rs}. Figure 4.2 visualizes such variations by scaling the property P^1, as it correlates to the magnitude of property degradation. The temperature range of the transition process remains constant for all P^1 variations. The transition temperature T^{rs} defines the temperature domain in which the transition occurs and represents a characteristic property of each bond type. Its variation leads to an effective temperature-shift of the transition and slightly changes the progression, i.e. depending on the temperature ratio of Eq. 4.3. The Weibull parameter m^{rs} characterizes the temperature range and the characteristic of a transition. Values above one lead to the typical S-shape, which becomes sharper with increasing magnitude of m^{rs}, while values below 1 lead to an asymptotic behavior. The magnitude of m^{rs} depends on the bond type, material homogeneity, and molecular structure (e.g. crystallinity). A

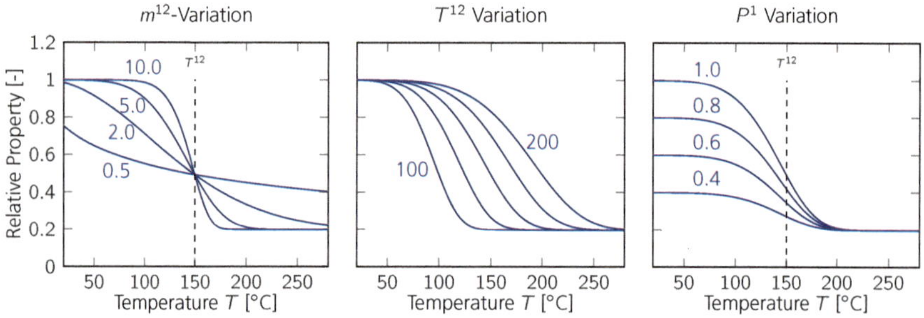

Fig. 4.2 Parameter sensitivity study for variations of a two state Mahieux model representing a single transition process; Reference parameters: $P^1 = 1.0$, $P^2 = 0.2$, $T^{12} = 150\,°C$, $m^{12} = 5.0$

low value indicates a restricted molecular movement and a wide temperature range for the transition, while a high value of m^{rs} indicates few impediments and an amorphous structure, resulting in a sharp onset of degradation. Mahieux demonstrated that the amount of filler material in polybutadiene leads to lower degradation slopes, which correlates inversely with the Weibull coefficient m^{rs} [16]. Similar behavior is observed in the amount of crystallinity in thermoplastic materials affecting the glass transition behavior [15]. The model has been extended to include plasticity by Reifsnider [17], and has been applied to reproduce the frequency dependency of the storage modulus for PMMA by Boumbimba et al. [19]. Further model utilizations are described in [4, 11, 14]. In conclusion, the mathematical approach outlined above provides the following characteristic points for a single transition model, which is useful in the process of fitting the model to experimental data.

$$P\left(T=-\infty\right) \rightarrow P^{r} \tag{4.4}$$

$$P\left(T=\infty\right) \rightarrow P^{s} \tag{4.5}$$

$$P\left(T=T^{rs}\right) = \frac{1}{e} \approx 0.368 \tag{4.6}$$

Gibson's publication [20] builds upon the previously described model of Mahieux. It aims on investigating the mechanical laminate behavior under fire loading. Although Mahieux's model provides satisfactory results, Gibson proposes an alternative mathematical formulation based on the hyperbolic tanh function. This approach has been successfully used to model the behavior of elastic properties, tensile strength, and compressive strength over temperature. Feih et al. [21, 22] employed the material model to simulate the mechanical behavior in a one-sided fire case and showed certain validity compared to experimental results. The approach has also been applied to predict mechanical failure under fire loads for different composite material systems [23, 24].

The similarity of Gibson's and Mahieux's approaches provides comparable courses of mechanical property over temperature. Therefore, the extensive work of the team around Mouritz, Gibson, and Feih [20–27] is valuable, as it demonstrates the applicability and validity of such models for elastic and strength properties of FRP composites. This is relevant as the foundation of these approaches is based on the polymer matrix system. The comparison of both approaches further indicate, that the distribution term could potentially be replaced by a different functional approach. Such tailored adaption of individual terms could help to describe the bond dissociation procedure more accurately. However, this is a theoretical consideration for hypothetical, specific improvements.

4.2.3 Model Definition

The previous paragraph explains selected approaches to model mechanical properties as function of temperature. In the present work, the approach of Mahieux is used, as shown in Eq. 4.3, as it considers the phenomenological nature of the material to describe the temperature relation of its macroscopic properties. The approach allows for modelling each property individually and is therefore useful to capture the transversal isotropic behavior of the UD ply. Its physically-based nature limits the regressive variability within data fitting and introduces a certain reliability to the existing property scatter over temperature, which is evident from experimental characterization. The final material models are derived using manual fitting and considering a certain range of property domains known from the constitutive behavior, such as the glass transition temperature. Further discussions on this topic are included in the following sections.

The material models are specifically developed for the tCFRP composite material and the selected mathematical formulation of Mahieux's approach is a 3-state version (2-transition steps), as shown in Eq. 4.7. The superscripts a, b, and c correspond to the relaxed material states and the superscripts ab and bc define the transitions between those states. In this regard, ab represents the „weak"transition within the glassy state, designated by β^{CE}, while bc corresponds to the glass transition of the PES and is described by α^{PES}.

$$\hat{P}_{ijk}(T) = \left(P^a_{ijk} - P^b_{ijk}\right) \cdot exp\left(-\left(\frac{T}{T^{ab}}\right)^{m^{ab}}\right) + \left(P^b_{ijk} - P^c_{ijk}\right) \cdot exp\left(-\left(\frac{T}{T^{bc}}\right)^{m^{bc}}\right) + P^c_{ijk} \tag{4.7}$$

Material properties that refer to the model are designated in the following using the $\hat{(\cdot)}$ symbol, as shown in Eq. 4.7 for the general property $\hat{P}_{ijk}$. The model parameters T^{ab}, m^{ab} and T^{bc}, m^{bc} define the transition characteristic.

The modelling is done for the five moduli of the UD lamina and its six strength properties:

- Elastic properties: $P_{ijk} \in \{E_{11t}, E_{22t}, E_{11c}, E_{22c}, G_{12}\}$
- Strength properties: $P_{ijk} \in \{R_{11t}, R_{22t}, R_{11c}, R_{22c}, R_{12}, R_{13}\}$

It is worth noting that the Poisson's ratio and the CTE's are assumed to be independent of temperature. The valid modeling temperature domain is defined by 23 °C to 250 °C. This range is defined based on the characterization temperature domain to avoid extrapolation. Tests performed at higher temperature are additional training points and serve to improve the model accuracy.

The determination of model parameters is crucial as they directly impact model accuracy. To derive a reliable and accurate model, a robust database must be used. However, the number of data points obtained from the static characterization is limited. Therefore, the model determination process, i.e. parameter fitting, is conducted in two steps.

In the initial investigation, outlined in Sect. 4.3.1, the model parameters T^{ab}, m^{ab} and T^{bc}, m^{bc} are derived from the DMA measurements. The DMA data provides a large number of measurement points, enabling a good establishment of the temperature-dependent property progression. However, as the DMA measurement technique applies only low mechanical forces, the resulting behavior is not fully representative for higher load domains. Therefore, in the second step, these model properties are applied for modelling the elastic and strength properties, using the static characterization results, but may be adapted if substantially improving the model's accuracy. The relaxed model state parameters, P^a_{ijk}, P^b_{ijk} and P^c_{ijk}, are manually fitted for the individual property database and in reference to Eqs. 4.4 and 4.5.

When transferring material parameters from DMA to the elastic and strength properties, it is important to acknowledge the phenomenological nature of the model approach. The approach describes the kinetic correlation between the energy level and bond dissociation. As a result, in a combined thermomechanical load case, both the thermal state and the mechanical load state are sensitive to the final model parameters. At a low mechanical load state, more thermal energy is required to break the chains, while at a highly loaded mechanical state, less thermal energy is needed, meaning a lower temperature is required for the same level of bond dissociation. This leads to a degradation of properties over temperature from a mechanical standpoint. Due to the higher mechanical load state in static characterization, the elastic and strength property transitions occur at lower temperatures compared to the DMA model basis, although the transition progression is similar and model parameters are remained. This reflection is only considered for the translation of model parameters to the characterization results. An additional correlation between load state and temperature is not introduced and linear elasticity is assumed.

4.2.4 *Uncertainty and Error Modelling*

The experimental characterization provides a database consisting of material property values at each test temperature T^{tp}. An inherent uncertainty of the material property exists at each of these temperature points and is assumed here to occur in normal statistical distribution (see Sects. 3.5 and 4.1). During the determination of model parameters, the mean material property values μ_P are used. However, a deviation between the apparently measured mean value and the true mean property value can exist as a consequence to the limited number of specimen tested, which is known as statistical correlation. In addition, material properties determined through the model and or obtained by experiment may not match exactly. Such deviation can negatively affect the load-carrying capability of the final product. Therefore, it has to be considered in the final material property design allowable. The deviation between model and measurement is designated here as model uncertainty.[5] This section describes a procedure to derive property allowables that considers these uncertainties.

The model approach of Mahieux can be utilized to determine such model uncertainty. The difference between the experimental points μ_P and the related model points $\hat{\mu}_P$ defines the model uncertainty and is determined at each test temperature point. To compare uncertainties over the full temperature range and domain of property degradation, it is more appropriate to use a Relative Error (RE) quantity, which is calculated as the fraction of model values and experiment points:

$$\hat{\underline{e}}_P^{RE}\left(T_{tp}\right) = \frac{\hat{\underline{\mu}}_P\left(T_{tp}\right)}{\underline{\mu}_P\left(T_{tp}\right)} - 1 \tag{4.8}$$

A RE of zero indicates coincidence of model and experiment. Positive values, $\hat{e}_P^{RE} > 0$, refer to model values greater than measured (overestimation), while negative values, $\hat{e}_P^{RE} < 0$, represent lower model values compared to the measurement (underestimation). Underestimation can be treated as a conservative model attribute. Overestimated property regions are critical, since they might cause preemptive failure. It is important to note that while underestimation of a material property may be considered a conservative model attribute in some cases, it may not always be the case for the variety of structures and load cases. This is because an underestimation of a material property could result in unexpected failure modes or reduced overall structural integrity, depending on the specific design and operating conditions. However, such cases are not further investigated here.

Because of the mentioned severity of preemptive failure, overestimating model uncertainties must be considered within the final property allowable. Once the model uncertainty is known, the material allowable over temperature can be determined by simple superposition of model and model uncertainty. Hence, the value of model uncertainty considered can be seen as a related MS. However, because of the low number of samples, the statistical nature of the problem is challenging. Due to the statistical correlation of the experimental training points in Eq. 4.8, it is unknown whether the model underestimates or overestimates the true property values and to what extent. This can also vary between the temperature points and complicates the determination of a model error function that would lead to an accurate property formulation. Due to the critical consequences associated with overestimation, a conservative representation of the model uncertainty is necessary and a simple procedure using a constant value of $\hat{e}_P^{RE}$ is employed. To be conservative, the most critical error obtained at the point of highest overestimation is used. It is determined by:

$$\hat{e}_P^{RE} = max\left(\hat{\underline{e}}_P^{+}\right) \quad \text{with} \quad \hat{\underline{e}}_P^{+} = \{\hat{\underline{e}}_P^{RE} \mid > 0\} \tag{4.9}$$

[5] Using the term "model error" may be incorrect since the phenomenology correctly displays the effective behavior. Therefore, „model uncertainty"may be a more appropriate term.

Using this approach to define the applied model uncertainty considers the most severe error to scale the related MS. This has a potential disadvantage where outliers can dominate this margin. It is therefore important that the overall temperature dependent property relation is well represented by the selected model function.

The approach described above can be utilized using different error metrics, while the RE is assigned here. However, since material properties decrease continuously with increasing temperature, the use of absolute error quantities is inadequate. Thereby, the highest error is likely to occur in the region of high property magnitude, causing a strong knock-down effect in the region of lower property magnitude where absolute errors might be lower. This approach limits the potential of the material performance in the high temperature domain but is conservative. On the other hand, RE quantities, as used in Eq. 4.8, cause an opposite discrepancy that results in property underestimation at low temperatures but leads to more accurate representations at elevated temperatures. Averaging error metrics, such as Root Mean Square Error (RMSE) or Root Mean Square Relative Error (RMSRE), are not entirely suitable due to their smoothing effect that overestimates the highest error regions. As the elevated temperature regime is the focus of this thesis, the RE metric is used.

Based on the explanation above, a conservative model function of the mean material property over temperature is obtained by the superposition of model mean and model uncertainty determined by Eq. 4.9. However, structural design requires consideration of material uncertainty as well. Based on the results discussed in Sect. 3.5 (see also Table 3.4), the final temperature-dependent property allowable is determined according to:

$$\hat{P}^{Alw}(T) = \hat{\mu}_P(T) \cdot \underbrace{\left(1 - \hat{e}_P^{RE}\right)}_{model} \cdot \underbrace{\left(1 - z \cdot \bar{cv}_P\right)}_{material} \tag{4.10}$$

The property model $\hat{P}_{ijk}(T)$ obtained by Eq. 4.7 is determined using the property mean values and expressed here by the term $\hat{\mu}_P(T)$ to highlight the applied statistical basis. It is multiplied by two effective Knock-Down Factor (KDF). The first accounts for the model uncertainty and the second for the material uncertainty. The formulation of $z \cdot \bar{cv}_P$ allows to consider confidence intervals related to the normal distribution which are used to determine *A-value* and *B-value* formulations, respectively, with $\bar{cv}_P$ being the average coefficient of variation listed in Table 3.4. The following four model representations are used through this work:

- Mean model: $\hat{e}_P^{RE} = 0, z = 0$
- Minimum mean model: $\hat{e}_P^{RE} \geq 0, z = 0$
- B-value: $\hat{e}_P^{RE} \geq 0, z \approx 1.645$
- A-value: $\hat{e}_P^{RE} \geq 0, z \approx 2.576$

Figure 4.3 provides an overview of the individual steps involved in determining the temperature-dependent property allowable, using a specific temperature T^i as an example training point for the model. Firstly, the statistical distribution of the property is determined (black, rectangle with error bars representing mean and standard

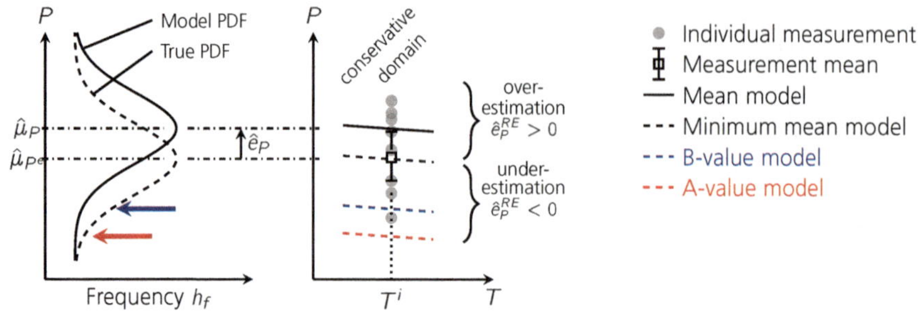

Fig. 4.3 Overview to the different model representations with respect to their statistical relation for the example of an overestimated training point

deviation) from a set of individual measurements (gray dots). Based on this distribution, the model is derived, which is visible as the black solid line, including a certain model error. Assuming that the error shown in the figure satisfies Eq. 4.9, the resulting minimum mean model function progresses exactly through this point, as highlighted by the dashed line. If another training point of the property set provides the model uncertainty quantity, the curve would be below and indicate conservatism at the shown point. In parallel, the PDF of the property is shifted accordingly. Finally, the blue and red dashed lines represent the allowable properties in terms of the *A-value* and *B-value*, respectively.

As mentioned earlier, it is important to acknowledge that the model mean and measurement mean are not exact, and a certain discrepancy in terms of the model error always occurs. Since the present characterization program tested a relatively low sample number, it is possible that such errors are relatively high compared to a test program with a high number of repetitions. However, even in such cases, a model error will exist, although it might be lower. Therefore, further investigations are required to study the model error. This includes the use of different metrics to quantify the error and to understand its implications for conservatism. Such studies could benefit from datasets with a higher number of temperature points and would therefore complement further model applicability studies.

4.3 Modelling of Allowable Properties for the Unidirectional Layer

This section covers the modelling of elastic and strength properties for the investigated tCFRP composite material. Sections 4.3.1 to 4.3.3 present the details of the model parameter derivation (i.e. model fitting) for each property and discuss the resulting behavior. Finally, Sect. 4.5 offers a detailed discussion of the material models in their entity and their applicability for predicting the mechanical behavior of the tCFRP composite material laminates over a range of temperatures.

Various error metrics are available to evaluate the accuracy of a model. In the context of material modelling, it is crucial to identify outliers and differentiate them from common uncertainties. Therefore, a combination of local and global error metrics is employed. The relative model error, $\hat{e}_P^{RE}$ (as shown in Eq. 4.8) is utilized to determine the maximum relative error (Maximum relative error (MaxRE)) as the highest local deviation.

$$MaxRE = max\left(\mid \hat{\underline{e}}_P^{RE} \mid\right) \tag{4.11}$$

Botchkarev [28] and Freund [29] propose global error metrics that enable an overall evaluation of model accuracy. In addition to the MaxRE, the RMSRE is commonly used as it provides an average of all deviations and can be used in comparison with the MaxRE.

$$RMSRE = \sqrt{\frac{1}{n}\sum_{i}^{n}\left(\hat{e}_{P,i}^{RE}\right)^2} \tag{4.12}$$

Comparing these error metrics allows for a more comprehensive assessment of the model accuracy. The MaxRE identifies potential outliers, while the RMSRE provides an average of all deviations and allows to compare against the natural variation of the material.

4.3.1 Parameter Determination Using DMA Measurements

The DMA measurement results are explained in Sect. 3.4.1 and Fig. 3.1. This section utilizes the collected data to derive the material model parameters. At first, the model parameters are determined for the neat resin. At second, these parameters are utilized to model the composite behavior for verification.

Figure 4.4 presents the comparison of the storage modulus obtained from the measurements (blue) and the models (black) for the neat CE polymer and the tCE polymer. The measured temperature range of the neat CE material is wider than that of the composite, enabling the glass transition of the CE material (α^{CE}) to be captured. Hence, the model can be derived for a higher range of temperature. As a result, the model of the neat CE comprises two transitions (β^{CE}, α^{CE}), and that of the neat tCE consists of three transitions (β^{CE}, α^{PES}, α^{CE}). The transition temperatures are chosen based on the material specifications (conf. Section 3.2) and Eq. 4.6, and the other model parameters are selected in accordance with the experimental data. Table 4.3 lists the corresponding model parameters.

Both models predict the material behavior with qualitatively good precision. For the CE material, the poorest accuracy occurs as underestimation at around 200 °C. For the tCE material, the highest errors are within the α_{PES} regime, where the model partially overestimates the measurement. However, this can be attributed to a poor parameter selection. Such curve sections, e.g. of highly changing property

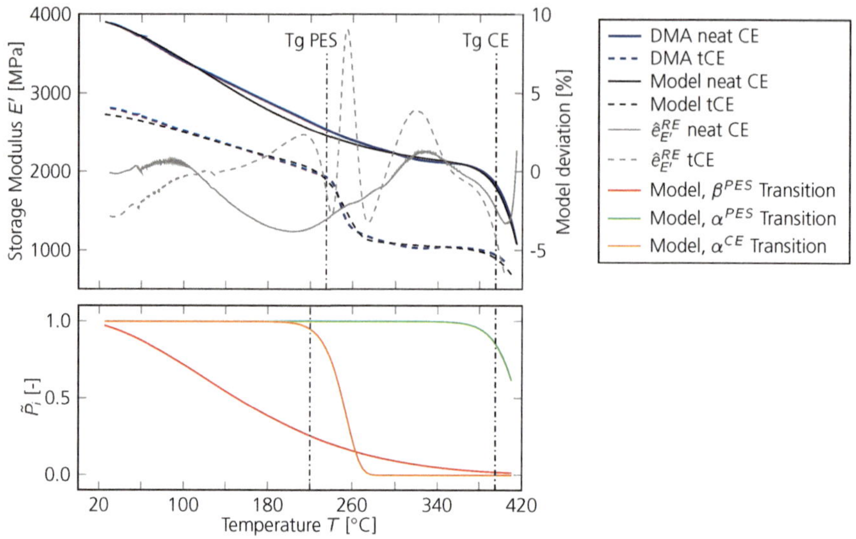

Fig. 4.4 Top: Comparison of DMA measurements (blue lines) to related material models (black lines) for the neat CE system (solid lines) and the tCE system (dashed lines). The model error is highlighted in gray color. The bottom part shows the model terms describing the individual transitions

rate, are vulnerable for errors as result to propagating material uncertainties. The consideration of these errors for allowable formulation is not further addressed here, since only sensitivities are studied. Overall, the accuracy of both models is good, with RMSRE values of $\pm 2.0\,\%$ and $\pm 2.4\,\%$ for the CE and tCE materials, respectively. Such errors are within the range of common standard deviations.

In the lower part of Fig. 4.4, the progression of the individual transitions' is emphasized. Each transition term is drawn individually and normalized to its RT value. The β^{CE} mechanism (red) decreases in a continuous form and with a relatively low rate, seemingly following a hyperbolic behavior towards zero. It accounts for the overall decreasing property trend. Both glass transition curves (orange and green) transition within a comparable small temperature range of about 40 K and are responsible for relatively the spontaneous drop at a specific temperature.

Figure 4.5 compares models and measurements for the composite materials in the fiber direction (left) and transverse direction (right). The temperature dependency of these composites differs from that of the polymer materials. Models in red dashed lines use the parameters determined previously for the neat polymers. The comparison of these models to the measurements indicates a certain mismatch in the elevated temperature regions, requiring adaptation. The deviation observed can be attributed to the addition of the fibers and their relatively constant properties, as well as to the interaction between the fibers and the matrix. In the fiber direction, the sensitivity of property degradation to the α^{PES} transition is low due to the fiber-dominated character of the property. The property degradation is initially noticeable when approach-

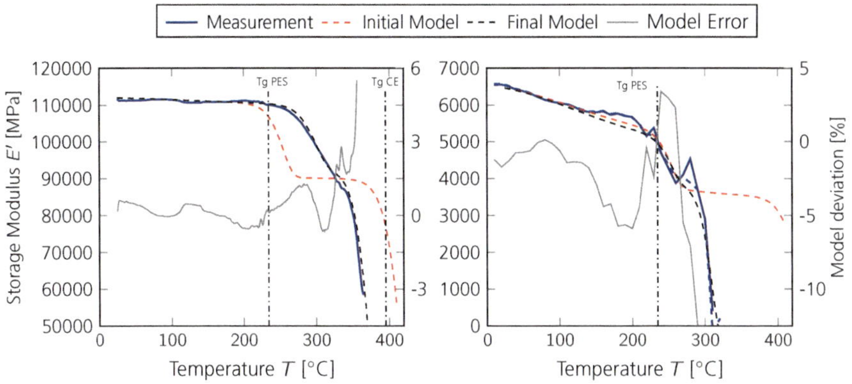

Fig. 4.5 Comparison of UD composite DMA measurements (solid, blue lines) to related material models (dashed lines) in fiber direction (left) and in transverse direction (right). Material models in red lines are based on model parameters determined for the neat resin and material models in black lines use adapted parameters. The model error refers to the model with adapted parameters and is shown in gray color

ing approximately 270 °C. At higher temperatures, the storage modulus decreases rapidly, which can be interpreted as severe consequence of matrix softening leading to loss of fiber support and related load transfer capability reduction.

The storage modulus in transverse direction shows good correlation between model and measurement for the low temperatur domain and the α^{PES} transition range. For transverse loading, the specimen are damaged when reaching 300 °C, indicating similar interfacial problems. This temperature domain is slightly below that observed in fiber direction and highlights the criticality of matrix softening and interfacial behavior for transverse composite properties.

To address these deviations, the model transition temperatures are adapted while the model parameters remain constant, conf. Table 4.1. The adapted models are depicted by the black dashed lines. Both models show good overall accuracy with RMSRE values of $\pm 0.5\,\%$ and $\pm 2.7\,\%$ for the longitudinal and transverse storage moduli, respectively. The RE remain within a certain range of approximately $\pm 5\,\%$ until 350 °C for the longitudinal storage modulus and 270 °C for the transverse storage modulus. The intended temperature range of 250 °C shows reasonable accuracy.

4.3.2 *Elastic Property Modelling*

This section describes the material models developed for elastic properties. Figure 4.6 presents the comparison of each property model to the measurement data over temperature. The models are determined following the prescribed procdure. In contrast to the DMA models, the α^{PES} transition temperature is reduced to $T^{bc} = 235\,°C$

Table 4.1 Overview to model values determined based on DMA measurements using a model with 4 states and 3 transitions

Material	β^{CE}		α^{PES}		α^{CE}		Model errors			
	m^{ab}	T^{ab}	m^{bc}	T^{bc}	m^{cd}	T^{cd}	Range	$\hat{e}^{RE}$	RMSRE	MaxRE
	[–]	[° C]	[–]	[° C]	[–]	[° C]	[° C]	[%]	[%]	[%]
CE	1.8	180	–	–	30	420	T <400	1.3	2.0	3.8
tCE	1.8	180	20	255	30	420	T <400	8.3	2.4	8.3
tCFRP, long.	1.8	180	20	300	30	375	T <300	1.3	0.5	1.3
tCFRP, transv.	1.8	180	20	255	30	310	T <270	6.8	2.7	10.2

to account for the transition onset at lower temperature. This is expected as a result to the higher mechanical loads considered here and has been mentioned before. Table 4.2 provides a summary of the applied model parameters and the resulting errors, which are further discussed below.

Elastic Properties in Fiber Direction

The simplification of a constant property behavior in dependence of temperature is used to model the moduli in the fiber direction. This assumption is supported by the experimental results obtained from the material characterization. The proposed model approach can represent such behavior by setting the transition degradation magnitude H^{rs} to zero (see Eq. 4.3). This formulation indicates that the property is insensitive to the related polymer transition, and the model equation becomes independent of the model parameters. In a heterogeneous material, the property would degrade by the amount of resin degradation, while the fiber is assumed to behave constantly. As the dimension of such neat polymer degradation is within the range of property variation, a constant formulation is applicable in this case.

The model's representation of the tensile modulus is relatively accurate, with an RMSRE of 1.2 % and a MaxRE of 1.6 %. These values are below the material's standard deviation and no significant outliers are observed. For the compressive modulus in fiber direction, the increasing property trend at elevated temperature deviates from the model. The MaxRE of 13.2 % occurring at 250 °C leads to a relatively high RMSRE of 6.6 %. Assuming that the property increase is caused by test deficiencies allows to omit this single 250 °C training point for error evaluation. In this case, the resulting RMSRE of 2.7 % is within the range of standard deviation and the MaxRE is equal to the standard deviation.

In both cases, all individual measurement results fall within the *B-value* domain. However, due to the small number of test repetitions, this is only a partial view of the true property scatter.

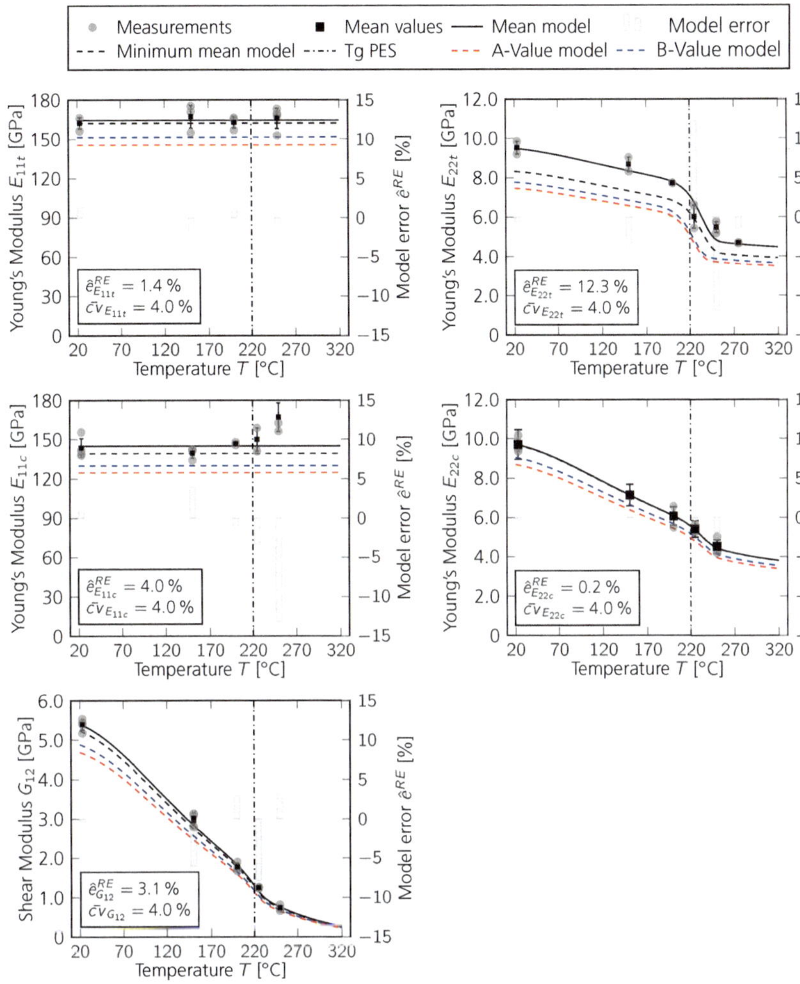

Fig. 4.6 Material models (lines) of elastic properties compared to measured values (points). Solid lines represent the mean model and dashed lines consider KDF according to Eq. 4.10. Gray circles and black squares represent individual and mean measurement results, respectively. Gray error bars indicate the model error between mean value and mean model

Elastic Properties for Transverse Direction and Shear Loading

The relaxed model properties of the moduli in transverse direction are almost equal at RT, $E^a_{22t} \approx E^a_{22c}$, conf. Table 4.2. The different degradation rates of tension and compression are scaled based on the relaxed state model properties, e.g. $E^b_{22t} > E^b_{22c}$. For transverse tension, this results in a property remaining at a higher magnitude before changing into a steeper slope towards T_g. The overall fit of the mean tension model is achieved with an RMSRE of 7.2 %, which is slightly higher compared to the related property variation coefficient. The MaxRE of 12.3 % occurs at 225 °C

Table 4.2 Overview to model parameters applied for elastic properties and resulting relative errors. Models are based on 3 states and 2 transitions and use the constant parameters: $m_{ab} = 1.8, m_{bc} = 20$. The values in braces refer to a reduced dataset used for error evaluation, as described within the text

Sign	P^a [MPa]	T^{ab} [°C]	P^b [MPa]	T^{bc} [°C]	P^c [MPa]	$\bar{cv}$ [%]	$\hat{e}^{RE}$ [%]	RMSRE [%]	MaxRE [%]
E_{11t}	164495.6	185.0	164495.6	235.0	164495.6	4.0	1.4	1.2	1.6
E_{22t}	9524.0	185.0	7143.0	235.0	4300.0	4.0	12.3	7.2	12.3
E_{11c}	144885.3	185.0	144885.3	235.0	144885.3	4.0	4.0	6.6 (2.7)	13.2 (4.0)
E_{22c}	9800.0	185.0	4410.4	235.0	3430.3	4.0	0.2	0.9	1.8
G_{12}	5500.0	175.0	475.0	220.0	5.5	4.0	3.1	4.8	7.9

and a similar error magnitude is found at 250 °C but with different sign. These errors are sensitive to the overall model accuracy. The deviations occur within the range of the α^{PES} transition. The overestimating error at 225 °C is critical and therefore used as maximum model error. It is observed that both allowable models progress clearly below the lowest measured value, which could indicate that the model uncertainty (given by the mentioned overestimation) is resulting in a potentially too high knock-down.

The transverse compression modulus decreases with a higher rate at low temperature compared to the transverse tensile modulus. The model shows a smaller rate change throughout the α^{PES} transition and is relatively accurate, with a MaxRE of 1.8 % and an RMSRE below 1 %. All individual measurements are within the range of the allowable models. However, some measurements approach the margin of the *A-value* basis, indicating that a higher sample rate would increase confidence.

The shear modulus degrades with a higher rate compared to the previously discussed properties and shows an almost linear trend over the full temperature domain. To model such behavior with good accuracy, the model temperature parameters, T^{ab} and T^{bc}, are decreased, see Table 4.2. Such lower transition temperature could be a result to the load conditions in shear, which are critical for the bond breakage. The resulting errors of RMSRE and MaxRE are determined by 4.8 % and 7.9 %, respectively. The most critical overestimating error occurs at 250 °C. The individual measurements are within the determined *B-value* allowable basis, indicating that the considered model error is relatively conservative.

4.3.3 Strength Property Modelling

Figure 4.7 illustrates the comparison between model and experimental values for the strength properties, using equal denotations as before. White circles are used in addition to indicate assumed values. These values are necessary for modelling due

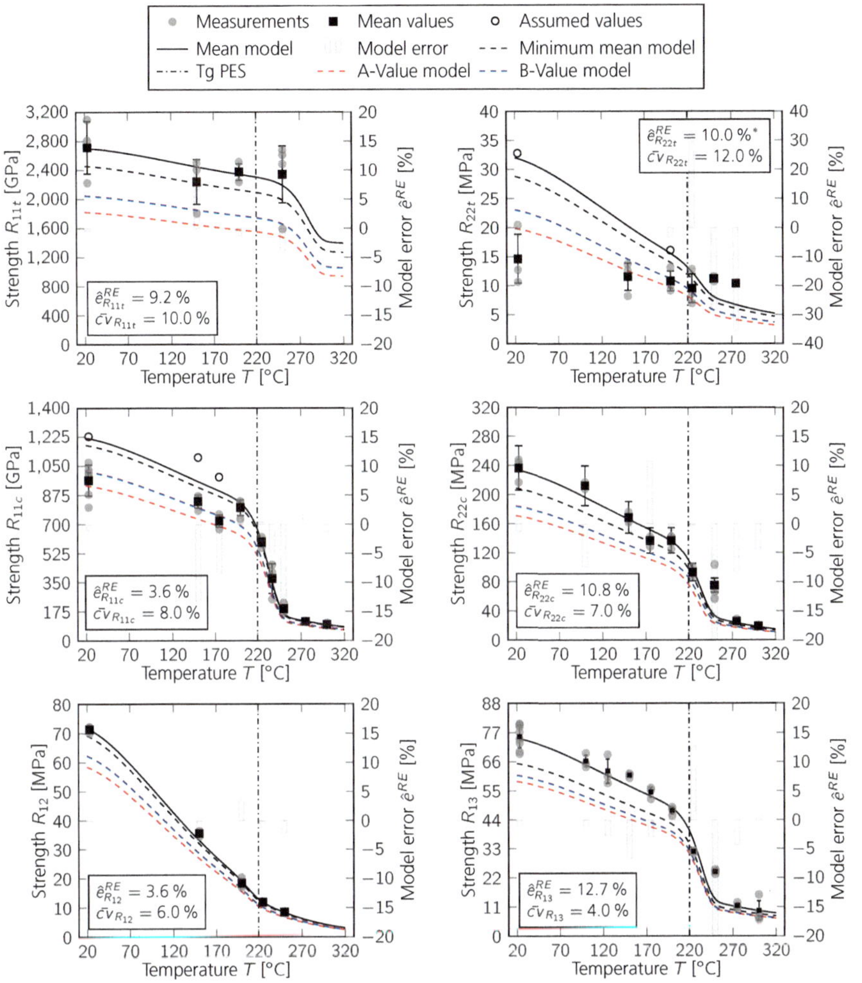

Fig. 4.7 Material models (lines) of strength properties compared to measured values (points). Solid lines represent the mean model and dashed lines consider KDF according to Eq. 4.10. Gray circles and black squares represent individual and mean measurement results, respectively. White circles represent assumed values in cases where invalid failure modes are observed (conf. Chap. 3). Gray error bars indicate the model error between mean value and mean model

to the presence of invalid failure modes observed during transverse tensile testing[6] and compression strength testing in fiber direction.[7] Table 4.3 provides an overview to the model parameters and the resulting errors.

[6] The assumed values represent individual measurement results selected from batch 1 of transverse tensile tests, as described in Sect. 3.4.2.

[7] The assumed values are determined by adding a constant shift to the measured values, conf. Sect. 3.4.4.

Table 4.3 Overview to model parameters applied for strength properties and resulting relative errors. Models are based on 3 states and 2 transitions and use the constant parameters: $m_{ab} = 1.8$, $m_{bc} = 20$. The values in braces refer to a reduced dataset omitting 250 °C values and used for error evaluation, as described within the text. * marks values of deficient basis for purpose of comparison

Sign	P_a [MPa]	T_{ab} [°C]	P_b [MPa]	T_{bc} [°C]	P_c [MPa]	$\bar{cv}$ [%]	$\hat{e}^{RE}$ [%]	RMSRE [%]	MaxRE [%]
R_{11t}	2712.0	185.0	2169.6	280.0	1356.0	10.0	9.2	5.1	9.2
R_{22t}	32.4	185.0	8.0	235.0	3.5	12.0	29.7*	24.3*	35.5*
R_{11c}	1228.0	185.0	680.0	235.0	45.0	8.0	3.6	7.4	14.4
R_{22c}	236.3	185.0	94.5	235.0	5.0	7.0	10.8	20.0 (5.9)	57.7 (10.8)
R_{12}	73.3	160.0	5.0	210.0	1.0	6.0	3.6	2.0	3.6
R_{13}	75.4	185.0	37.7	235.0	6.0	4.0	12.7	14.5 (5.3)	43.1 (12.7)

Fiber Direction

The tensile strength in the fiber direction exhibits no significant degradation within the temperature range of the experimental data. The variation of mean properties over temperature is similar to that obtained by the standard deviation of the material ($\approx \pm 10.0\,\%$). However, it is likely that the R_{11t} value degrades with temperature due to reduced fiber support capability caused by matrix degradation and reduced interfacial properties. However, this degradation is not statistically verified here. Nevertheless, such degradation becomes relevant within the critical load domain, where loads must be transferred to adjacent regions when individual fibers break. If the matrix or interface properties are reduced, this corrupts such mechanisms and leads to strength reduction. To account for this, a certain temperature-induced strength degradation is applied through the model using the selected model parameters. Hence, the model follows an average path of the experiment points and the trend of individual measurements located at the lower property bound follows the overall trend of the models. The low sensitivity of the matrix to the property can also be translated to the α^{PES} transition characteristic. It is assumed to occur at higher temperatures compared to the matrix-dominated properties, and is selected based on the longitudinal DMA measurements by $T^{bc} = 280$ °C. The selection of the relaxed properties beyond this transition (P^c) has no experimental evidence. Additional measurements at elevated temperature would be needed for a more reliable selection of these parameters (T^{bc}, P^c). However, this is not relevant for the current investigation, as the temperature domain of interest is limited to 250 °C. This exercise highlights the capability of the model for possible extrapolation or to introduce margins of safety by comprehensive parameter selection.

The model error falls within the range of standard deviation and is equal to the MaxRE. The RMSRE is even below and indicates a good model accuracy. Individual measurement results fall between A-value and B-value models and indicate a suitable model uncertainty selection close to optimum.

The compressive strength in fiber direction is more sensitive to temperature compared to the tensile strength. The α^{PES} transition causes a spontaneous degradation due to the related matrix softening process, as explained in the experiment Sect. 3.4.4. The failure mode changes in this temperature range from a brittle to a soft fracture pattern. It is assumed that the measured strength values are still representing the maximum load carrying capacity of the material at such conditions and are therefore conservative margins. The values can therefore be used as model input. However, this can not be verified and further investigation is required to address the changing failure modes observed in this study, and to gain a more comprehensive understanding of the effect to the overall failure behavior of MD laminates.

The mean model agrees well with the experimental data at RT and for $T > 200\,°\mathrm{C}$. The latter represents the first measurement point used for model fitting. However, the two assumed values at temperatures below this range are underestimated by up to 15.6 %, resulting in the MaxRE. Interestingly, the two related mean measurement points are relatively close to the mean model. This indicates that the reason for the invalid failure mode could correlate with temperature, such as residual stresses relax towards higher temperature. The highest overestimation, i.e. $\hat{e}^{RE}$, occurs within the transition domain at $T > 237\,°\mathrm{C}$. The overall accuracy of the mean model, with an RMSRE of about 7.4 %, is still within the range of the variation coefficient. The deviation of the assumed values affect the RMSRE negatively. A better effective model accuracy can therefore be expected at the elevated temperature domain. Despite the invalid values at RT, all individual measurements remain within the allowable range, while some values at high temperature are relatively close to the A-value model.

Transverse Direction and Shear

The modelling of the transverse tension strength is significantly affected by test deficiencies. This leads to using a selected individual measurement result and assumptions as basis for model fitting. For temperatures up to 200 °C, the two assumed points (shown as white circles) are used, while for temperatures above, mean properties are considered. Despite the measured properties above 225 °C are relatively constant, it is clear that a certain degradation must take place. Therefore, the model properties are selected with a focus on conservative behavior and underestimation. Due to the poor quality of the dataset, it is not possible to derive a reliable model and to conduct a suitable discussion of model errors. A model error of 10.0 % is applied within the modelling procedure leading to the allowable models. Further work is needed to improve the quality and quantity of the experimental dataset and to further discuss material modelling for the transverse tension strength.

The transverse compression characterization exhibits valid failure modes at all temperatures. The mean properties display a consistently decreasing behavior. Similar to the compression strength in the fiber direction, the rate of degradation changes significantly around the α^{PES} transition. The mean model shows a good qualitative correlation. Quantitatively, the two highest model errors occur at 175 °C and 250 °C. The former overestimates the measured value, resulting in a $\hat{e}^{RE}$ value of 10.8 %. The latter represents the MaxRE and is of conservative nature, as it underestimates the measured property. It is comparably high and thus sensitive to the model RMSRE. If this latter value is omitted, the RMSRE drops to 5.9 %, which is below the material standard deviation. Overall, all individual measurements are within the allowable range of the *B-value*. This indicates a relatively conservative model error selection.

The IPS strength exhibits a similar degradation behavior to the shear modulus. A steady decrease in property is observed from RT onwards, and no distinct drop around the α^{PES} transition is observed. The mean model shows relatively good accuracy, with a MaxRE of 3.6 %, which equals the critical model error, and an RMSRE of 2.0 %. Both are below the variation coefficient.

In contrast to the IPS strength, the ILSS follows a similar trend as observed for the other properties, with a smooth degradation at low temperature that changes to a higher slope across the α^{PES} transition. The highest model errors occur within the transition temperature domain. The comparably high MaxRE of about 43.1 % is obtained at 250 °C, where the model underestimates the measurement. The critical model error is determined at 225 °C. Although the overall mean model accuracy is good, the RMSRE is sensitive to the high MaxRE value. It is determined by 14.5 %, or 5.3 % if omitting the point of MaxRE. The minimum mean model describes a relatively high knock-down compared to the mean model. Mostly all individual measurement points are located above this curve, while the allowable models are even below. The onset of plastic failure is observed at a temperature domain equal to the highest model errors and it is unclear if the used failure loads represent a conservative margin in this domain. Hence, such failure is indicated as invalid failure mode regarding test standards. Therefore, the allowable models determined by Eq. 4.10 do not represent a conservative model in this domain. If the model domain would be reduced to a lower temperature of valid failure, e.g. $\leq$ 200 °C, the related model errors would lead to a lower knock-down between mean model and minimum mean model. Although the ILSS models are not relevant for structural analysis of this work, this example highlights the importance of a concise investigation of the database. Discussing test and model results and their validity is essential to ensure reliable and accurate modeling of the material behavior. In this case, further studies would be needed to investigate how to treat the changing failure mode and the related onset of plasticity for a conservative property determination, e.g. by adapted experimental data evaluation methods that provides data for training the models.

4.4 Stiffness Behavior of Multidirectional Laminates in Dependence of Temperature

The captured material models provide insights into the behavior of the UD tCFRP material as a function of temperature, highlighting the orthotropic behavior that results in different load directions from the contribution of fiber and matrix properties and their alignment. In practice, MD laminates, consisting of stacked UD plies with different orientations ϕ_i, are more commonly used than purely UD laminates. The elastic behavior of an MD laminate arises from the interaction between its constituent parts, and its effective properties are crucial for structural design. The engineering properties of the laminate can be described using the CLT [30], which condenses the properties of individual plies into those of a single homogenized composite material. The resulting effective properties can be expressed using the ABD matrix [30]. The ABD matrix is defined by the extended Hooke's law, which relates normal forces $\underline{N}$ and bending moments $\underline{M}$ to laminate strains $\underline{\varepsilon}$ and curvatures $\underline{\kappa}$:

$$\begin{pmatrix} \underline{N} \\ \underline{M} \end{pmatrix} = \begin{bmatrix} \underline{\underline{A}} & \underline{\underline{B}} \\ \underline{\underline{B}} & \underline{\underline{D}} \end{bmatrix} \begin{pmatrix} \underline{\varepsilon} \\ \underline{\kappa} \end{pmatrix} \tag{4.13}$$

with the membrane stiffness matrix $\underline{\underline{A}}$, the bending stiffness matrix $\underline{\underline{D}}$, and the coupling stiffness matrix $\underline{\underline{B}}$. The latter relates normal forces to curvature and bending moments to strains. It is zero if the laminate is symmetric [30] and all layers have equal properties, compensating for individual ply deformations relative to the laminate symmetry axes. However, due to the temperature-dependent behavior of elastic properties, temperature gradients can cause an effective $\underline{\underline{B}}$ that differs from zero even for symmetric laminates. Although this effect is not studied in detail in this work, it highlights the complexity that arises for distributed temperatures. The terms of the $\underline{\underline{A}}$, $\underline{\underline{B}}$, and $\underline{\underline{D}}$ matrices can be calculated using the plane-stress reduced stiffness matrix $\underline{\underline{Q}}$, and taking into account the local ply orientation via transformation condition.

$$\underline{\underline{\tilde{Q}}}_k = \underline{\underline{T}}(\phi_k)\, \underline{\underline{Q}}_k\, \underline{\underline{T}}^T \tag{4.14}$$

For the plane-stress state, the $\underline{\underline{Q}}$ terms are defined according to [30]:

$$Q_{11} = \frac{E_{11}}{1-\nu_{12}\nu_{21}}, \quad Q_{22} = \frac{E_{22}}{1-\nu_{12}\nu_{21}}, \quad Q_{12} = \frac{\nu_{12}E_{22}}{1-\nu_{12}\nu_{21}}, \quad Q_{66} = G \tag{4.15}$$

The terms of the stiffness matrices are directly related to the elastic properties of the UD plies and therefore lead to similar relations to temperature. The stiffness matrices of the laminate can be determined by integrating over all layers k:

$$A_{ij}(T) = \int_{-\frac{t}{2}}^{\frac{t}{2}} \tilde{Q}_{ij}(T,\phi)\, dz = \sum_{k=1}^{n} \tilde{Q}_{ij,k}(T)\,(z_k - z_{k-1}) \tag{4.16}$$

$$B_{ij}(T) = \int_{-\frac{t}{2}}^{\frac{t}{2}} \tilde{Q}_{ij}(T,\phi)\, z dz = \frac{1}{2}\sum_{k=1}^{n} \tilde{Q}_{ij,k}(T)\left(z_k^2 - z_{k-1}^2\right) \tag{4.17}$$

$$D_{ij}(T) = \int_{-\frac{t}{2}}^{\frac{t}{2}} \tilde{Q}_{ij}(T,\phi)\, z^2 dz = \frac{1}{3}\sum_{k=1}^{n} \tilde{Q}_{ij,k}(T)\left(z_k^3 - z_{k-1}^3\right) \tag{4.18}$$

Table 4.4 Engineering constants at RT conditions for different laminate setups

Designation	Laminate setup	E_{xt} [MPa]	E_{yt} [MPa]	E_{xc} [MPa]	E_{yc} [MPa]	G_{xy} [MPa]	ν_{yx} [–]
FI	[0]	164496	9481	144885	9703	5400	0.32
MAP	$[0/\pm45]_s$	92771	25844	82767	24018	23813	0.73
CP	$[0/90]_s$	87398	87398	77702	77702	5399	0.03
QI	$[0/\pm45/90]_s$	72106	72106	64352	64352	17675	0.20
TR	[90]	9481	164496	9703	144885	2700	0.02

Equations 4.16 to 4.18 show that a temperature gradient over thickness affects the effective laminate properties. To compare the effective laminate properties under a pure membrane stress state, the engineering constants of the laminate are used [30]:

$$\tilde{E}_x = \frac{1}{t_{la}}\left(A_{11} - \frac{A_{12}^2}{A_{22}}\right), \quad \tilde{E}_y = \frac{1}{t_{la}}\left(A_{22} - \frac{A_{12}^2}{A_{11}}\right), \quad \tilde{E}_{xy} = \frac{1}{t_{la}A_{66}}, \quad \tilde{\nu}_{yx} = -\frac{A_{12}}{A_{11}} \tag{4.19}$$

Table 4.4 and Fig. 4.8 present a comparison of different exemplary laminate configurations. Figure 4.8 illustrates the axial engineering modulus (in the x-direction) normalized to its RT value and the orthotropy ratio, defined as the ratio between the axial and lateral engineering moduli. Tensile and compressive loads are indicated by solid and dashed lines, respectively. The comparisons are performed on five different laminate setups, ranging from the extreme laminates using fiber direction plies only (FI) and transverse plies only (TR) to a laminate with a preferential direction in the axial direction (MAP), a symmetric CP laminate stacking, and a quasi-isotropic (QI) laminate setup.

The behavior of the FI and TR UD laminates over temperature is similar to their corresponding UD properties, with the FI laminate being constant over temperature and the TR laminate exhibiting significant degradation with increasing temperature. It should be noted that compressive loads and tensile loads have different effects on TR laminates (and transverse UD properties) highlighted via the normalized curves. The three "intermediate" laminates (MAP, CP, and QI) show a similar but diverging trend, with significantly lower differences between tensile and compressive loads. The overall degradation magnitude is also lower compared to the TR laminate, which is a result of the fiber dominating character of 0° -plies. The constant fiber direction properties of 0° -plies result in curves that progress almost parallel to 1, with small degradation magnitudes for the MAP, CP and QI laminates below 10 % and a smooth property drop within the investigated temperature range. The position of the laminate curve between the FI and TR laminates is determined by the number and ratios of integrated plies and their orientation, and this behavior is observed similarly over the full temperature domain. This behavior is in agreement with the principles of CLT and highlights the sensitivity of 90° plies to the elastic behavior within a MD laminate, particularly at high temperatures.

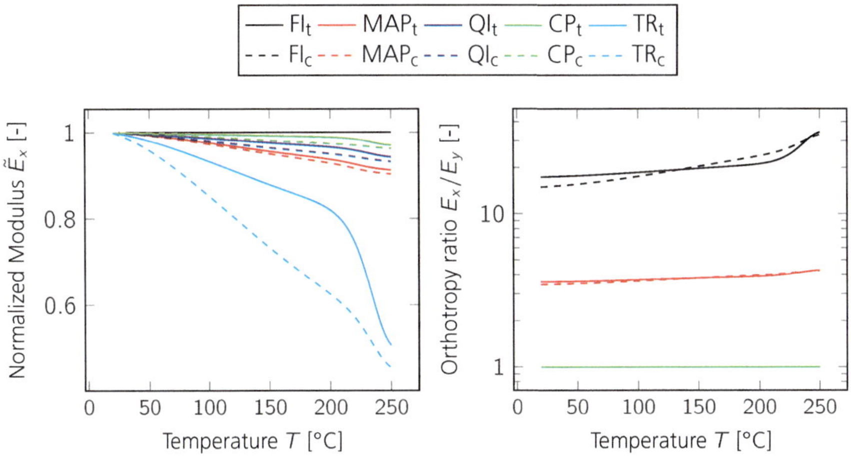

Fig. 4.8 Exemplaric comparison of the effective elastic behavior for different MD laminates, conf. Table 4.4. Solid lines represent tensile loads and dashed lines indicate compressive loads

The orthotropy ratio indicates the relation between the longitudinal stiffness and the lateral stiffness of the laminate. It is constant over temperature and equal to 1 for the QI and CP laminates due to the symmetry condition, which includes identical ply orientations within both laminate directions. The orthotropy ratio of the MAP laminate is between that of the FI and QI laminates, as the ±45° plies provide further stiffness in transverse direction. For compressive loads, the ratio changes from 3.4 at RT to 4.1 at 225 °C, and further increases at elevated temperature.

4.5 Assessment and Discussion

This chapter aims to investigate the first working hypothesis, which forms the foundation for the overall research objective. As explained in the previous sections, a modelling procedure is developed to determine the orthotropic material properties of the UD lamina as a function of temperature. These models describe the elastic and strength components of the material and are suitable for being utilized in FEA. Critical errors are determined where the model overestimates the measurement. Such errors are used to define a KDF which considers the model uncertainty in the final allowable models. Figure 4.9 presents a comparison of the material and model uncertainties, determined throughout material characterization (conf. Sect. 3.5) and material modelling, respectively. In most cases, the model uncertainty is lower compared to the material uncertainty, indicating good model quality. Therefore, the first working hypothesis can be confirmed since the resulting allowable models are conservative. However, in some cases, the material uncertainty is above the materials standard deviation, indicating poor model quality and creating exceptions to this, as

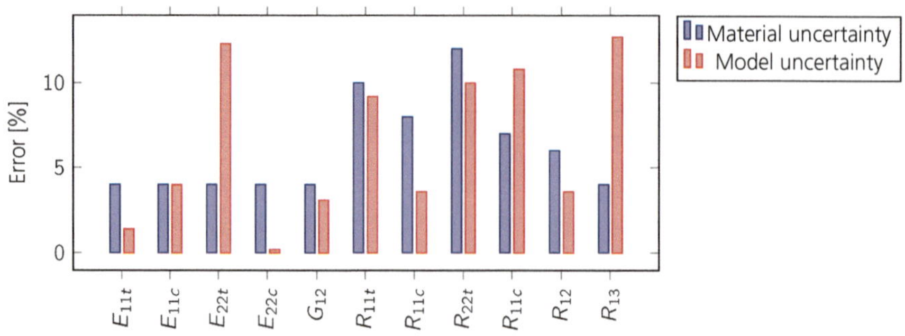

Fig. 4.9 Comparison of experimental material uncertainties and model uncertainties

explained earlier. This is especially the case for transverse tensile loading, where testing showed severed problems. Further R_{11c} and R_{13} present cases, where the overall model was relatively accurate but the highest critical error, i.e. the model uncertainty was caused in the domain of PES T_g. A more detailed view of these points and possible improvements are discussed below.

The largest errors are typically observed within the α^{PES} transition region, which occurs between 225 °C and 250 °C. There could be several reasons and solutions to this. A refined parameters selection could reduce errors, e.g., a less steep slope in the transition domain may improve the model accuracy for multiple properties. Another possibility is that all model parameters could be selected based solely on minimizing errors, and for each property individually and without considering the governing phenomenological nature of the approach. Such an pure optimization approach would resemble a typical regression or surrogate modelling technique and may not fully capture the physics of the system. The use of such a procedure can be particularly inaccurate, especially when dealing with a small and scattered database, as in the present case.

Notable deviations from the expected model parameters are observed in the shear properties. Such deviation could be caused by preemptive polymer bond relaxation compared to the other properties or can be a result related to the stressing of matrix and interface. However, this interpretation is based on the fitted parameters and requires further physical verification. A possible enhancement would be an appropriate factor that describes the sensitivity of matrix and interface stressing on the final property, e.g. in dependence on the load direction.

A model that is built on scattering data always contains model errors, and the challenge is to ensure that it provides a conservative formulation. The methodology used to introduce these model uncertainties works sufficiently and provides a conservative model representations for all properties. However, the methodology sometimes leads to overconservative models in specific temperature spaces. This is especially because of scatterous data and insufficient statistical representation. A poor model parameter selection could also cause such deviations but is certainly linked to the noise of the data. This underscores the importance of having a qualitatively and quantitatively robust dataset to derive accurate and reliable material models.

To address this issue, three general approaches can be taken. First, increasing the number of samples in the dataset can provide a better statistical representation of the data and allow for adjustments to the model parameters. This is a fundamental recommendation for improving model accuracy. Second, limiting the temperature domain of the model to a specific range can avoid considering higher model errors outside this domain, although it restricts the material's usable service temperature range. For example, in the present case, using only the temperature domain up to 200 °C would significantly reduce errors and still be beyond typical service temperature limits. Finally, applying a temperature factor on the transition temperature as another form of conservatism can be useful. This approach would primarily impact the transition region and shift the onset of the property drop to lower temperatures, reducing the maximum (critical) property errors. This would allow for higher accuracy of the material's properties at low temperatures while still ensuring its applicability at elevated temperatures. However, further investigation is needed to explore the feasibility and effectiveness of these approaches.

Another potential and related enhancement of the model is to integrate a temperature-dependent material uncertainty formulation. This would require a suitable dataset and would also enable further refinement of the model error definition.

4.6 Summary

A temperature-dependent formulation of material properties is essential for analyzing the structural behavior in distributed thermal conditions. This chapter describes the development of methods that are useful to derive allowable material property models depending on temperature. Such material models are derived for the mechanical properties of the UD tCFRP material solution characterized in Chap. 3. These models provide conservative estimates of the property magnitudes and can be utilized within structural analysis and sizing of applications exposed to thermomechanical conditions. This result confirms working hypothesis 1 and aligns with the overall research objective. The models are applied within the following Chapters using the *A-value* margin.

The selected material modelling approach, proposed by Mahieux, is originally developed to describe the temperature dependent properties of polymers but is also applicable to FRP materials. This type of model is employed because of its phenomenological nature which allows the property relation over temperature to be obtained based on the characteristic behavior of the matrix material. Hence, it is particularly useful for datasets with a low number of temperature points. By utilizing the matrix temperature dependency and scaling it appropriately, the individual properties of the composite materials can be estimated. This allows for the consideration of the specific characteristics of each material property, such as whether it is fiber or matrix dominated, and to respect the different magnitudes.

To ensure conservativity, the model approach is enhanced to consider material uncertainties and model uncertainties. It is crucial to consider the model uncertainties, as overestimation could lead to catastrophic failures. However, the determination of these uncertainties highly depends on the quality and quantity of the dataset since the nature of material properties is statistical. The use of these models for the investigated material provides valuable insights. The resulting property allowables, which incorporate material uncertainty and model uncertainty, are conservative, but they can be partially overconservative, resulting in unexploited load-carrying capacity. The overconservatism is mainly due to the low quantity of available data, leading to individual outliers that cause high model uncertainty, ultimately resulting in a knock-down of the property allowable. To address these issues, enhancements of the model, limitations of the modelling space, and alternative types of conservatism are proposed for the formulation of accurate allowables and efficient property characterization programs.

References

1. C. G'sell, D. Jacques, J.P. Favre, Plastic behaviour under simple shear of thermosetting resins for fibre composite matrices. J. Mater. Sci. **25**(4), 2004–2010 (1990). https://doi.org/10.1007/BF01045756
2. C.X. Sun, K.J. Yoon, Characterization of elastic-plastic behavior of AS4/PEEK thermo-plastic composite for temperature variation. J. Compos. Mater. **25**(10), 1297–1313 (1991). https://doi.org/10.1177/002199839102501003
3. M.F. Ashby, *Engineering Materials 2: An Introduction to Microstructures and Processing*. International Series on Materials Science and Technology, 4th edn. (Butterworth-Heinemann, Oxford, 2012)
4. C.A. Mahieux, K.L. Reifsnider, Property modeling across transition temperatures in polymers: a robust stiffness-temperature model. Polymer **8**(42), 3281–3291 (2001). https://doi.org/10.1023/A:1011230620286
5. Y. Bai, T. Vallée, T. Keller, Modeling of thermal responses for FRP composites under elevated and high temperatures. Compos. Sci. Technol. **68**(1), 47–56 (2008). ISSN: 02663538. https://doi.org/10.1016/j.compscitech.2007.05.039
6. Y. Bai, T. Keller, T. Vallée, Modeling of stiffness of FRP composites under elevated and high temperatures. Compos. Sci. Technol. **68**(15–16), 3099–3106 (2008). ISSN: 02663538. https://doi.org/10.1016/j.compscitech.2008.07.005
7. Y. Bai, T. Keller, Modeling of mechanical response of FRP composites in fire. Compos. Part Appl. Sci. Manuf. **40**(6–7), 731–738 (2009). ISSN: 1359835X. https://doi.org/10.1016/j.compositesa.2009.03.003
8. B. Yu, V. Till, K. Thomas, Modeling of thermo-physical properties for FRP composites under elevated and high temperature. Compos. Sci. Technol. **67**(15), 3098–3109 (2007). ISSN: 0266-3538. https://doi.org/10.1016/j.compscitech.2007.04.019. https://www.sciencedirect.com/science/article/pii/S0266353807001996
9. Y. Bai, T. Keller, Modeling of strength degradation for fiber-reinforced polymer composites in fire. J. Compos. Mater. **43**(21), 2371–2385 (2009). ISSN: 0021-9983. https://doi.org/10.1177/0021998309344642
10. N. Platzer, Progress in polymer engineering. Ind. Eng. Chem. **61**(5), 10–30 (1969). https://doi.org/10.1021/ie50713a003. https://pubs.acs.org/doi/pdf/10.1021/ie50713a003
11. C.A. Mahieux, A systematic stiffness-temperature model for polymers and applications the prediction of composite behavior. Dissertation. Blacksburg, VA: Virginia Tech, 1999

12. C.A. Mahieux, K.L. Reifsnider, Influence of temperature on the durability of polymer composite systems, in *ICCM 12 Proceedings* (1999)
13. C.A. Mahieux, K.L. Reifsnider, S.W. Case, Property modeling across transition temperatures in PMC's: Part I. Tensile properties. Appl. Compos. Mater. **8**, 217–234 (2001). https://doi.org/10.1023/A:1011282704357
14. C.A. Mahieux, K.L. Reifsnider, Property modeling across transition temperatures in PMC's: Part II. Stress rupture in end-loaded bending. Appl. Compos. Mater. **8**, 235–248 (2001). https://doi.org/10.1023/A:1011230620286
15. C.A. Mahieux, K.L. Reifsnider, Property modeling across transition temperatures in polymers: application to thermoplastic systems. J. Mater. Sci. **37**, 911–920 (2002). https://doi.org/10.1023/A:1014383427444
16. C.A. Mahieux, K.L. Reifsnider, Property modeling across transition temperatures in polymers: application to filled and unfilled polybutadiene. J. Elastomers & Plast. **34**(1), 79–89 (2002). ISSN: 0095-2443. https://doi.org/10.1106/009524402022348
17. K.L. Reifsnider, V. Tamuzs, S. Ogihara, On nonlinear behavior in brittle heterogeneous materials. Compos. Sci. Technol. **66**(14), 2473–2478 (2006). ISSN: 02663538. https://doi.org/10.1016/j.compscitech.2006.04.003
18. R.F. Boyer, Dependence of mechanical properties on molecular motion in polymers. Polym. Eng. Sci. **8**(3), 161–185 (1968). ISSN: 0032-3888. https://doi.org/10.1002/pen.760080302
19. R.M. Boumbimba et al., Dynamic mechanical properties of PMMA/organoclay nanocomposite: experiments and modeling. J. Eng. Mater. Technol. **133**(3), (2011). https://doi.org/10.1115/1.4004052
20. A.G. Gibson, Laminate theory analysis of composites under load in fire. J. Compos. Mater. **40**(7), 639–658 (2005). ISSN: 0021-9983. https://doi.org/10.1177/0021998305055543
21. S. Feih et al., Modelling the compression strength of polymer laminates in fire. Compos. Part Appl. Sci. Manuf. **38**(11), 2354–2365 (2007). ISSN: 1359835X. https://doi.org/10.1016/j.compositesa.2007.04.013
22. S. Feih et al., Modelling the tension and compression strengths of polymer laminates in fire. Compos. Sci. Technol. **67**(3–4), 551–564 (2007). ISSN: 02663538. https://doi.org/10.1016/j.compscitech.2006.07.038
23. A.G. Gibson et al., High temperature and fire behaviour of continuous glass fibre/polypropylene laminates. Compos. Part Appl. Sci. Manuf. **41**(9), 1219–1231 (2010). ISSN: 1359835X. https://doi.org/10.1016/j.compositesa.2010.05.004
24. A.G. Gibson et al., Modeling composite high temperature behavior and fire response under load. J. Compos. Mater. **46**(16), 2005–2022 (2011). ISSN: 0021-9983. https://doi.org/10.1177/0021998311429383
25. A.G. Gibson et al., The integrity of polymer composites during and after fire. J. Compos. Mater. **38**(15), 1283–1307 (2004). ISSN: 0021-9983. https://doi.org/10.1177/0021998304042733
26. A.P. Mouritz et al., Review of fire structural modelling of polymer composites. Compos. Part Appl. Sci. Manuf. **40**(12), 1800–1814 (2009). ISSN: 1359835X. https://doi.org/10.1016/j.compositesa.2009.09.001
27. S. Feih, A.P. Mouritz, Tensile properties of carbon fibres and carbon fibre-polymer composites in fire. Compos. Part Appl. Sci. Manuf. **43**(5), 765–772 (2012). ISSN: 1359835X. https://doi.org/10.1016/j.compositesa.2011.06.016
28. A. Botchkarev, Evaluating performance of regression machine learning models using multiple error metrics in azure machine learning studio. SSRN Electron. J. (2018). https://doi.org/10.2139/ssrn.3177507
29. S. Freund, Metamodelle mehrerer Genauigkeitsstufen zur Stabilitätsanalyse versteifter Schalenaus Faserverbundwerkstoffen. Dissertation. TU Braunschweig, 2018. https://elib.dlr.de/137590/
30. C. Mittelstedt, W. Becker, Strukturmechanik ebener Laminate. 1. Auflage. Darmstadt, 2016. ISBN: 978-3-935868-99-0

Chapter 5
Methods for Structural Analysis at Distributed Temperatures

Section 2.3 and Fig. 2.6 provide an overview to the analysis procedure developed in this work which allows to consider distributed temperatures and is referred to as DiA procedure. It is a systematic approach that combines various analysis techniques to determine the mechanical structural behavior of composite structures under distributed thermal conditions and to assess the behavior regarding thermal load variation. This chapter describes all individual steps involved in this procedure and applies the allowable material property models determined in Chap. 4. The chapter concludes with describing an example of a DiA analysis run, which serves as the foundation for the following Fig. 2.6 where the procedure is applied.

5.1 Finite Element Analysis Procedure

Finite Element Analysis (FEA) is a widely used technique for analyzing the structural behavior of various materials and structures. In FEA, a structure is discretized into smaller elements. Analyzing the behavior of each element allows to account for a high level of detail and to handle complex structures and load cases. This work utilizes a set of sequential FEA to determine the mechanical behavior of structures under specific thermal conditions. This FEA process represents a single deterministic structural analysis and is part of the overall DiA procedure, as indicated in Fig. 2.6.

Figure 5.1 depicts the steps involved into this sequential FEA procedure. At the beginning of the procedure, the *Thermal Analysis* step determines the temperature distribution, $T(x, y, z)$, that results to predefined thermal load and Boundary Condition (BC). These conditions cause a thermal structural response and the resulting temperature distribution is the basis for the subsequent *Thermal Deformation Analysis* step. The sequential coupling of these two analysis methods does not account for

M. Liebisch, *Design Load Allowables for Composite Plates Exposed to Thermomechanical Loads*, Mechanics and Adaptronics,
https://doi.org/10.1007/978-3-031-71352-1_5

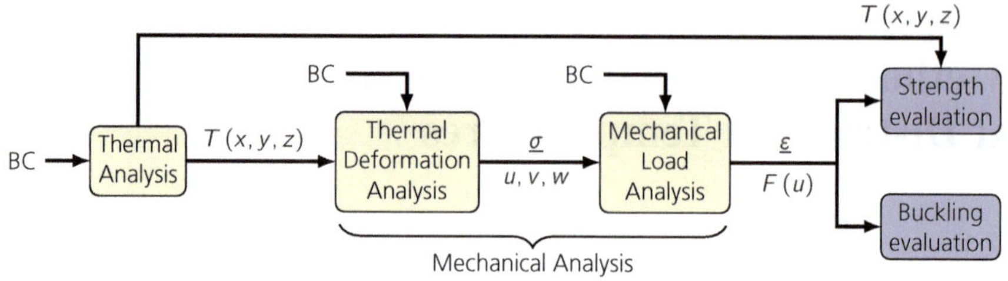

Fig. 5.1 Illustration of the sequential steps involved in a deterministic structural analysis process. Individual FEA steps are shown in bright yellow boxes. Postprocessing steps are shown in blue boxes

backward coupling, i.e. any mechanical analysis does not influence the temperature results of the thermal analysis.

The *Thermal Deformation Analysis* is the first part of the mechanical analysis. It calculates the state of deformation and stress that results to thermal expansion from an initial thermal state to the final temperature distribution while considering mechanical BC's. The material models developed in Chap. 4 are employed to account for the effective material properties that depend on the local temperature. Based on the deformation and stress state obtained, the *Mechanical Load Analysis* applies external mechanical loads and BC that are representing the mechanical load state. The behavior captured this analysis step refers to the combined thermomechanical load state, since the initial deformation represents the deformations caused by thermal loads. The strains $\underline{\varepsilon}$ describing the deformation of each element and the externally applied forces F of this analysis step are finally used to evaluate the structural behavior. In a postprocessing step, the buckling and material failure are assessed, in dependence of the temperature distribution, to obtain critical design values which are related to the thermal and mechanical loads applied through this series of analyses.

The consideration of thermal analysis and especially distributed load conditions is the innovation compared to the current uniform FEA (UnA) procedure. In UnA, a uniform temperature and constant material properties are assumed, resulting in simplified thermal deformations. However, the described approach (DiA) consist of temperature distributions leading to thermal deformations that depend on thermal gradients and material properties that vary with temperature. The following subsections explain the details of the individual FEA steps, including the applied BC's and the basic background needed to determine stresses in thermomechanical conditions. Procedures used for postprocessing are discussed in Sects. 5.2 and 5.3, respectively.

5.1.1 Thermal Analysis

The background of this thesis is on applications that are exposed to external heat sources, which cause local hotspots. Size and magnitude of these hotspots depend on the external boundary and load conditions, as well as on the setup of the structure. The DiA methodology enables engineers to account for such distributed temperatures by involving a thermal analysis step that considers thermal conditions corresponding to realistic conditions. The objective here is to reproduce the temperature conditions at the moment of severe mechanical loading rather than analyzing the transient behavior. In that regard, the DiA procedure considers varying temperature distributions to account for thermal load uncertainties, which may depend on time in a transient thermal load case or uncertainty in the boundary conditions.

The governing physical law for *Thermal Analysis* is the partial differential equation (PDE) of heat transport, which is applied to a specified system and represents a function that depends on the spatial coordinates and time. The equation considers all parameters of the system, e.g. the material properties, and its boundaries that influence thermal energy and calculates the resulting evolution of temperature over time. This PDE is related to the first law of thermodynamics, which requires energy equilibrium within closed systems [1] and the stationary state solution refers to the situation where the temperature does not change over time. Assuming the consistency of mass for a control volume V, i.e. an element, with surface areas A, the PDE of heat transport can be formulated as following [2].

$$\underbrace{-\nabla\left[\lambda\nabla T\right]+}_{conduction}\ \underbrace{\varrho c_p\frac{\delta T}{\delta t}}_{internalenergy} = \underbrace{Q^{int}+Q^{ext}}_{heatsources} \tag{5.1}$$

Equation 5.1 is crucial for understanding the thermal behavior of the component, as it comprises three types of terms. The left side terms describe the temperature distribution within the control volume (the structure of interest) and how it changes over time. It includes the heat capacity of the system and thermal conductivity in all directions. The right side terms describe the heat sources that add or remove heat from the control volume. Internal energies Q^{int} resulting from exothermic material reactions or mechanical deformation are not considered in Eq. 5.1 as they are neglected in this work. Other types of heat source are obtained by energy exchange with external sources, denoted by Q^{ext}. This term represents all forms of heat transfer through the surfaces of the component, including convection and radiative heat transport, which are temperature-dependent and sensitive to the internal and external conditions of the structure.

The distribution of external heating, i.e. the thermal load distribution, represents the source of the hotspot development within the structure. A temperature field which is spatially distributed within the x-y-plane is used to simulate the development of such hotspots. The temperature distribution applied to the heated side of a structure is expressed as T^{hot}.

$$T^{hot}(x,y) = T^{uni} + T^{grad}\cdot exp\left[-\left(b^x\cdot(x+dx)\right)^2-\left(b^y\cdot(y+dy)\right)^2\right] \tag{5.2}$$

The function is composed of two parts: a uniform temperature part T^{uni} which represents a minimal thermal load experienced globally, and a non-uniform part that generates the local thermal peak load responsible for the hotspot distribution. The local thermal peak load is created using a two-dimensional Gaussian function that generates a temperature shape corresponding to a normal distribution. The width of the distribution is determined by the shape parameters b^x and b^y, while the hotspot location is specified by dx and dy. The in-plane thermal gradient is considered using the factor T^{grad}, and the maximum hotspot temperature is obtained by summing the uniform and non-uniform temperature parts. Adapting these parameters allows to consider varying thermal conditions. Figure 5.2 presents exemplary temperature distributions generated using Eq. 5.2.

Figure 5.3 (left) illustrates a stiffened structure representing a part of a typical lightweight aircraft component and consisting of a single skin field (marked through red dashed line). As the external heat source represents a specifically distributed thermal load that is applied to a single surface of the structure, e.g. by applying Eq. 5.2, thermal gradients will develop in any direction of the part. When such structures are exposed to external heating, geometrical elements that are added to the skin (e.g. stringers) will impact the overall temperature distribution [3, 4]. A thermal analysis step is useful to analyze the detailed shape of temperature within the structure and depending on its geometrical details.

In the frame of the present work, a shell based modelling approach is used for FEA, as indicated within Fig. 5.3. The selection of this modelling strategy is based on the composite shell element selection of the subsequent mechanical analysis. These composite shell elements consider the laminate setup through related integration points distributed over thickness and in dependence on the integration rule used. The thermal analysis can be aligned to a similar definition of shell elements with thickness-wise section point distribution. As a result, the thermal FEA obtains a set of temperature degree of freedoms at any node representing the temperature values occuring over the thickness of the laminate. Consequently, thermal analysis allows to consider temperature gradients in thickness direction using shell elements. The right part of Fig. 5.3 visualizes this detail for an exemplaric heat transfer problem on a local spot of the structure. In this example, five black points are used to highlight

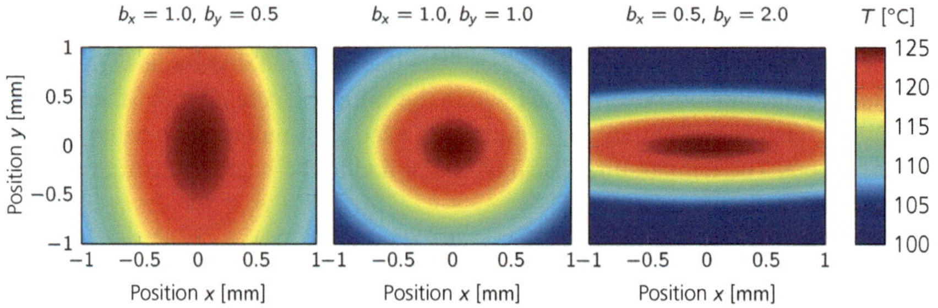

Fig. 5.2 Temperature distributions resulting to Eq. 5.2. $T^{uni} = 100\,°\text{C}$, $T^{grad} = 25\,\text{K}$

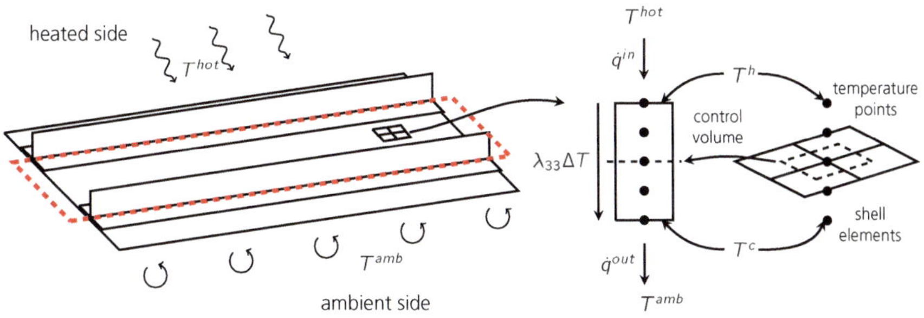

Fig. 5.3 Overview to applied thermal boundary conditions on the example of a stiffened panel (left). Concept of thermal analysis for a shell element approach (right). The skin field marked in red is focus of investigations in Chap. 6

the temperature points distributed over thickness by equal spacing and the thermal analysis applies the boundary conditions on both surfaces. Here, for instance, convection coefficients and flow temperatures are defined applying a certain heat flux density on each surface. The thermal analysis finally provides the equilibrium state that develops within the structure, i.e. the temperatures that result in all temperature points. The resulting surface temperatures T^h and T^c represent the maximum and minium temperature and define the thermal gradient that exists over thickness.

Variation of the thermal boundary conditions and the thermal load distribution leads to varying temperature distributions. Analyzing and considering these distributed temperatures is the scope of the present work and part of the DiA procedure. However, the selection of realistic thermal parameters and uncertainties is complex. Therefore, a detailed thermal analysis step as explained above is not considered for the investigations here (conf. Chap. 6) and is substituted by directly specifying the temperature distribution at the temperature points to obtain the input for mechanical analysis. The top part of Fig. 5.4 highlights exemplaric temperature distributions applied in analysis at the outer surfaces of the skin field test structure. This simplification leads to constant temperature gradients over thickness in a single deterministic analysis run. However, throughout the number of FEA considered in a DiA process, varying thickness-wise thermal gradients are used. Finally, this selection allows a better correlation between temperature distribution and mechanical behavior and allows to investigate the DiA procedure in comparison to uniform temperatures. Nevertheless, a detailed thermal analysis has to be considered in real world applications and especially for more complex structures.

5.1.2 Mechanical Analysis

The DiA procedure presented constitutes an analysis strategy that is, in general, applicable to structures subjected to thermomechanical loading. However, the focus

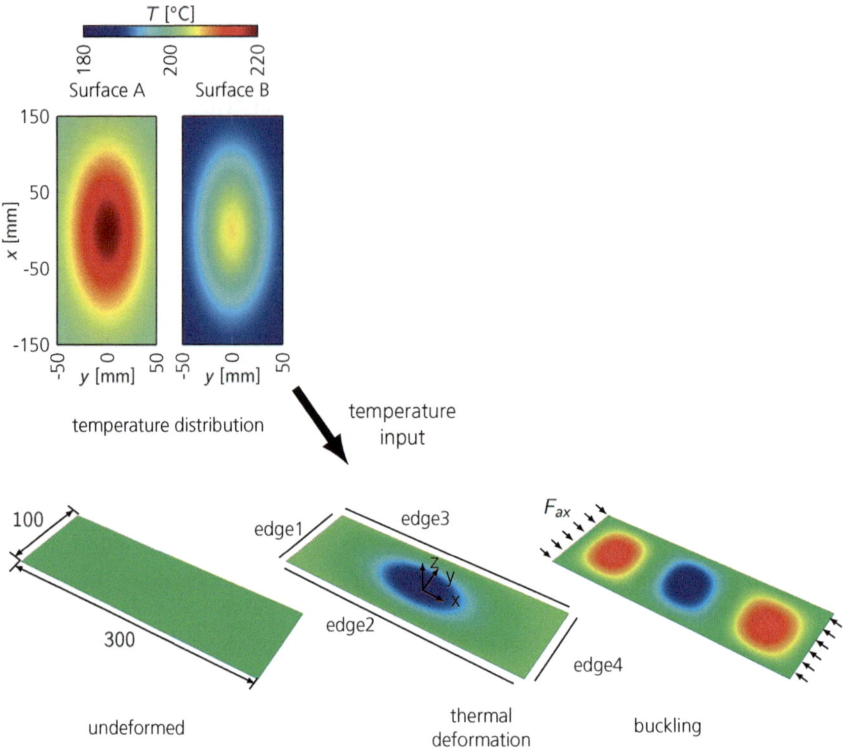

Fig. 5.4 Test structure and exemplary deformation resulting through mechanical analysis. The top chart highlights the temperature distribution resulting from *Thermal Analysis*. The mechanical analysis steps are highlighted at the bottom, from left to right: Initial state, deformation state after the *Thermal Deformation Analysis* step that applies the temperature distribuion and deformation state after the *Mechanical Load Analysis* step which includes axial compression loads

of this thesis is on lightweight composite aircraft structures. Such structures typically consist of thin skins that are stiffened by stringers and frames, as investigated in previous studies, for instance by Degenhardt et al. [5] or Wilckens [6]. To reduce the complexity of superposing structural mechanical effects that are introduced by temperature variations, a skin field is selected as the representative element. This skin field has a length of 300 mm and a width of 100 mm and is used as a test structure for the final investigations presented in Chap. 6 where also the selected laminate setups are explained. Figure 5.4 illustrates this skin field test structure and highlights the states considered in the mechanical analysis: the undeformed state (start of the analysis), the thermal deformation state resulting from the temperature distribution and determined within the *Thermal Deformation Analysis*, and the postbuckling shape resulting from the axial compression loads and analyzed within the *Mechanical Load Analysis*. Axial compression loads are investigated as they represent a critical mechanical load case under high-temperature conditions that depend on thermal deformation and the effective material property variations.

To analyze the mechanical behavior of the system, a shell-based approach, similar to the *Thermal Analysis*, is employed. This approach allows for the transfer of temperatures determined in the *Thermal Analysis* to the mechanical analysis model. The discretization used in the *Thermal Analysis* is inherited in the mechanical analysis, enabling consistent results. In the *Mechanical Analysis*, reduced linear shell elements are used to facilitate calculations. Material properties are assigned using a layerwise composite shell theory, which adopts the thickness-wise temperature information in each element and considers the temperature-dependent allowable properties derived in Chap. 4.

In the *Thermal Deformation Analysis*, the thermal expansion interacts with external constraints, causing the resulting deformation and stress state. Therefore, the definition of BC is essential as it scales the magnitude of thermal strain that translates into mechanical load and deformation. It is important to note that extreme BC, such as fully clamped or free conditions, do not represent realistic cases especially for thermomechanical load condtions, as discussed by Liebisch [3, 4]. Fully clamped conditions constrain any thermal expansion, leading to higher stress development. On the other hand, free boundary conditions allow the structure to deform without any stress development resulting from mechanical restraints. The realistic case must be somewhere in between. However, selecting representative BC can be challenging, especially if the stiffness and deformation of the global component are unknown. This is particularly relevant at thermomechanical conditions, as exterior structures also deform when exposed to temperature. Therefore, the mechanical analysis applied here, uses clamped conditions as they represent the most severe constraints. Regarding the investigations performed in Chap. 6, this choice is also motivated by the fact that it is simpler to analyze the behavior that occurs when BC are released rather than when they are reinforced. Hence, thermal expansion is constrained, and it translates into stress and distortion.

In the second analysis step, the *Mechanical Load Analysis*, the BC are retained but adapted to the axial compression load. Using a displacement controlled loading, compressive forces are applied by shifting the load edge (edge 1) towards the clamped edge (edge 4) while keeping both edges parallel. The Newton-Raphson technique involves an iterative procedure and is used to determine equilibrium states at incrementally increasing load states [6]. This allows to account for specific instabilities with temporary negative stiffness while applying the described displacement control. To ensure sufficient accuracy for the later analysis steps, a relatively small increment size is chosen since evaluating of the parameter output is done increment wise. A detailed overview to the analysis settings is provided within the Appendix Sect. C.1. The procedure used for stress evaluation is explained in the following Sect. 5.1.3, while the consideration of residual stress and the related presence of thermal stress is discussed in Sect. 5.1.4.

Geometrical imperfections are crucial for understanding the stability behavior of a structure. These imperfections represent a disturbance of the system and are used to trigger bifurcation in stability analysis. FEA allows to account for the presence of these imperfections using a specific option that shifts the nodes to a predefined, imperfect shape and thereby represents an artificial disturbance. Some studies use the phrase of geometrical imperfection to classify the variety of imperfections or system disturbances that can occur in structural stability analysis. Here, it simply refers to the mentioned analysis option. Thermal deformation and related stress represent a state of the structure that can, similarly, be seen as a disturbance of the perfect structure. As thermal deformations represents an actual response of the structure, it is more related to the final application and its environment than the nature of geometrical imperfections which refers to an inherent source of disturbance. By combining the use of both, geometrical imperfection assignment and thermal deformation analysis, a better understanding of how the imperfections affect the buckling behavior of the structure is gained. Furthermore, accounting for geometrical imperfections is essential to ensure comparability between the UnA reference scenario and the DiA, since the stability behavior of the reference scenario can be different. Section 5.2 and Chap. 6 provide more detailed information and results on this topic.

5.1.3 Stress Evaluation

The level of detail considered in a finite element model can significantly impact the computational cost of a FEA. In particular, the DiA procedure requires multiple FEA simulations to assess the final design values and its correlation to the temperature distribution, resulting in a significant computational effort. To address this issue, an efficient postprocessing methodology for determining stresses is presented in this section. This approach focuses on the governing effects of strain and stress development at thermomechanical load conditions, enabling the determination of stresses using only surface values of total strain and temperature. By reducing the number of measurement points required as FEA output, the overall analysis time can be significantly reduced, making it a valuable tool for design optimization and evaluation. The procedure explained below is implemented in a code outside the FEA software and simply uses the input and output information provided by the simulation.

The total strain, denoted by $\underline{\varepsilon}^{tot}$, is a physical quantity that characterizes the deformation of a material under external loads. It represents the measureable quantity of strain, as for instance measured by a strain gauge at the surface of a structure. In thermomechanical conditions, this measure contains the superposition of thermal strain $\underline{\varepsilon}^{th}$ and mechanical strain $\underline{\varepsilon}^{el}$. The total strain is therefore an important parameter in determining the resultant stress state of the material. Assuming that mechanical strains behave elastically and introducing Hooke's law, the total strain can be expressed as [7]:

$$\underline{\varepsilon}^{tot} = \underline{\varepsilon}^{el} + \underline{\varepsilon}^{th} = \underline{\underline{C}}^{-1} \cdot \underline{\sigma} + \underline{\varepsilon}^{th} \tag{5.3}$$

This expression is useful for explaining the strain and stress development occuring at combined thermomechanical loads. For example, in constrained BC, the thermal strain caused by a temperature change is restrained. The resulting mismatch between thermal strain and total strain translates into elastic strain and thus stress development. On the other hand, in fully relaxed BC, thermal strain is allowed to freely manifest and results in equal magnitudes of thermal strain and total strain, and therefore no stress development.

Equation 5.3 represents the governing relation within each element of FEA, with its BC determined by the behavior of adjacent elements or prescribed BC's. Therefore, it is possible to estimate the stresses within each ply from the total strains and temperatures at the laminate surfaces. Assuming linear strain distribution over thickness with z being the thickness coordinate, the total strain state of an element can be expressed as a superposition of membrane strain $\underline{\hat{\varepsilon}}_0$ and bending strain $\underline{\hat{\varepsilon}}_b$:

$$\underline{\hat{\varepsilon}}^{tot}(z) = z \cdot \underline{\hat{\varepsilon}}_b^{tot} + \underline{\hat{\varepsilon}}_0^{tot} \tag{5.4}$$

For small temperature gradients, it is also possible to approximate a linear relation of temperature over the thickness of a composite material. Assuming a constant, effective coefficient of thermal expansion (CTE) for the laminate, the membrane strain corresponds to the uniform temperature component, while thermal gradients result in bending terms. This relationship reveals how thermal gradients can cause bending and curvature and lead to out-of-plane deformation.

The relationships explained above, along with material properties that depend on temperature enable the calculation of laminate stress based solely on knowledge of total strains and temperature at the surfaces and can further be utilized for failure assessment. Figure 5.5 provides an example that illustrates this concept by showing the distributions of strain, temperature, and failure index over the thickness of a single element for a laminate exposed to combined thermomechanical loading.

The highlighted element experiences one-sided heating, resulting in a temperature gradient across the thickness t, as shown in the left chart. In addition, axial compression loads generate a strain state that comprises both membrane and bending strains. The second chart depicts the total strain in axial direction (x) of the laminate. To approximate temperature and strain as function over thickness, the assumptions of linear relations over thickness allows to reduce the number of measurement points needed to a minimum, i.e. the surface values shown by blue points. The third chart displays the thermal strains that occur due to temperature changes. The dashed red line represents the effective strain of the laminate, while the solid red line reveals the discrete magnitude of layerwise strains that result within the individual plies. This difference represents the inner restraint of a MD laminate and translates into mechanical strain and stress.[1] Such effect of different CTE's is often designated as thermal mismatch. The fourth chart illustrates the elastic strains resulting from

[1] This behavior is explained in more detail within the following Sect. 5.1.4 on the example of residual stress, conf. Fig. 5.6.

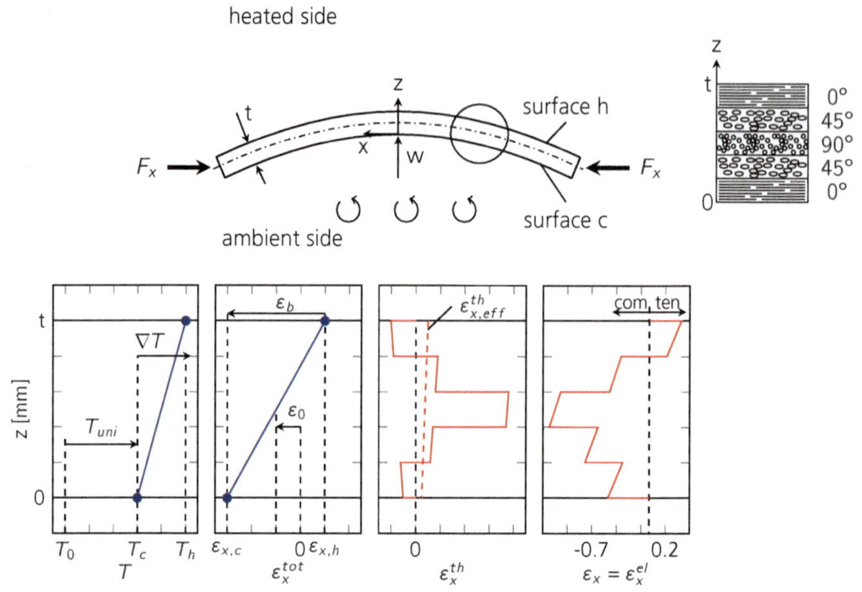

Fig. 5.5 Schematic explanation of the postprocessing procedure used for stress assessment. Blue markers represent the locations of FEA output used

Eq. 5.3, which are a superposition of elastic strains from thermal mismatch and external loading. These strains create stress development within the material. They are used to calculate stress for failure estimation.

Overall, the described approach provides an efficient and effective way to predict stress and conduct failure evaluation in laminates using only surface values of total strain and temperature. Within the scope of the present work, it is applied as a postprocessing step to reduce the overall analysis time. The basic simplifications made by this approach are obtained through the FEA considerations providing the data and the postprocessing steps explained above. An overview to the collection of assumptions is listed below.

- Linear-elastic material behavior
- Perfect connection between adjacent plies
- Plane state of stress
- First-order Shear Deformation Theory (FSDT): The cross section is straight over thickness but may not be normal to the central plane
- Neglection of transverse shear development
- Constant laminate thickness
- Linear relations of total strain and temperature over the laminate thickness.

5.1.4 Residual Stresses Developing from Composite Curing

The stress-free temperature is a crucial parameter for predicting the thermal deformation and stress development within a composite structure. Residual stress is a result of composite curing, and it develops in balance with deformation. Therefore, the stress-free state corresponds to the undeformed state [8]. By using the simplification of thermoelastic behavior, this state can be expressed by an equivalent temperature, which is designated as the stress-free temperature $T^{\sigma 0}$. Any temperature deviation from this state causes thermal strain and stress development, thereby contributing to the total loading of the part.

This section provides an overview of the considerations involved in residual stress development. In a sensitivity study, the residual stress development on the stress state is evaluated over temperature and for different laminates to highlight the importance. As part of this section, the magnitude of stress-free temperature is derived for the investigated composite material. This value is applied as input for stress assessment within the structural analysis procedure to consider its contribution to the overall loading state.

Estimation of Stress-Free Temperature

The process of composite curing involves the matrix transitioning from a viscous to a solid state and the development of connection to the fibers for being capable to transfer loads. However, this curing process also involves chemical reactions that cause the matrix to shrink, resulting in strain and stress in the final material [9]. After curing, the composite undergoes a cooling down process to ambient conditions which leads to the development of thermal strains. As a result, the residual state that occurs within a material is not only dependent on temperature but also on the chemical strains generated during curing. The coefficient of chemical shrinkage β and the degree of cure ξ quantify these resulting chemical strains. They act in analogy to the CTE and temperature, respectively, considering a change in magnitude to end up in strain. They are used in superposition similar to Eq. 5.3:

$$\underline{\varepsilon}^{tot} = \underline{\varepsilon}^{el} + \underline{\varepsilon}^{th} + \underline{\varepsilon}^{ch} = \underline{\underline{C}}^{-1} \cdot \underline{\sigma} + \underline{\alpha}\Delta T + \underline{\beta}\Delta\xi \tag{5.5}$$

Equation 5.5 allows to consider a stress-free condition in dependence on the superposition between thermal and chemical strains. Referring to the cure temperature avoids the presence of thermal strain, i.e. $\Delta T = 0$. Assuming thermoelastic material behavior and a constant CTE allows to approximate the stress-free temperature:

$$T^{\sigma 0} \sim T_{cure} + \frac{\beta \Delta\xi}{\alpha} \tag{5.6}$$

Since the chemical shrinkage is unknown here, Eq. 5.6 is only a simplistic visualization of the problem. The resulting chemical strain represents an inherent amount of strain in each composite material which contributes to residual stress. However, this simple approach neglects other effects. Valuable insights for defining the $T^{\sigma 0}$ can also be found within the literature.

Svanberg conducted a study using a thermoelastic material law to analyze the effects of temperature on residual stress and deformation state [9]. Other studies have utilized viscoelastic material models to account for the effects of time, temperature, and degree of cure. These studies, conducted by Hein [10], Hein et al. [11], Liebisch et al. [12, 13], have shown that residual stress is relatively insensitive to chemical resin shrinkage. Early studies by Hahn and Pagano neglected the contributions of residual stress from chemical strains due to the comparatively low stiffness assumed for the material [14]. However, viscoelastic material models have since been developed to consider stress relaxation, which is primarily sensitive at high temperatures and leads to lower residual stress. Weitsmann's study [15], for example, shows a stress relaxation of up to 20 % compared to a linear-elastic approach. Experimental studies determining the stress-free temperature using fiber Bragg grating sensors are performed by Prussak et al. [16] and Kappel et al. [17] for different fiber-metal-laminate composites. They indicate $T^{\sigma 0}$ to occur slightly below curing temperature. The study of Prasatya et al. [18] confirms similar observations for thermoset materials.

In the context of the cited literature, the stress-free condition is assumed to occur slightly below the cure temperature of 180 °C and is defined by a stress-free temperature of $T^{\sigma 0} = 160\,°\mathrm{C}$. This selection neglects contribution of resin shrinkage and accounts for a small amount of stress relaxation, leading to an effective stress-free temperature below the cure temperature. It is further assumed that the postcure processing does not contribute to the residual state. However, this selection is only a simplistic estimation based on practical considerations. It is useful for describing the sensitivities resulting to thermal strain, as part of the objective here and contributing to working hypothesis 3, conf. Sect. 2.2. A thorough investigation of this selection is recommended for the final use of this material.

Thermal Strain and Residual Stress Development of the Multidirectional Laminate

The previous paragraph explained the cause of residual stresses and the simplifications used to define the effective stress-free temperature. In the context of the present work, it is useful to predict the magnitude of residual stress that results in dependence of temperature. This knowledge is essential for studying sensitivities and gaining a comprehensive understanding of the system's behavior under combined thermal and mechanical loading. Therefore, it is necessary to formulate appropriate methods for predicting the magnitude of residual stress in laminates depending on temperature.

The effective properties of MD laminates differ from those of the individual UD plies, see Sect. 4.4. Hence, the internal constraints lead to stress development if the temperature deviates from the stress-free state. Figure 5.6 provides a schematic representation of the strains developed within a MD (cross-ply) laminate resulting purely from a change in temperature to a magnitude below the stress-free state. Due to the different CTE's of the plies in reference to the laminate coordinate system, the 0° ply will contract less than the 90° ply. The laminate, using the assumptions of CLT [7], shows an effective behavior that must be somewhere between the two.

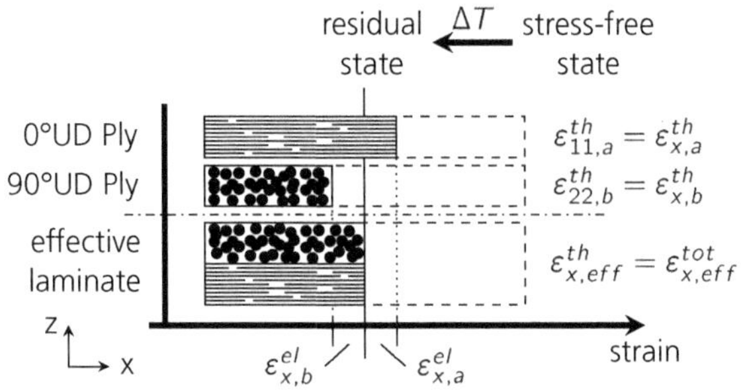

Fig. 5.6 Strain development in a MD laminate through temperature variation

Since no external mechanical loads are applied in this case, the thermal strain of the laminate equals its total strain. This total strain is the total strain that is applied to each ply. Therefore, a mismatch between the laminate strain and the individual thermal strain of the plies occurs. This margin is compensated by mechanical loading. Hence, it represents the elastic strain component of Eq. 5.5 and finally leads to stress development within each ply.
To calculate the resulting stress, one can transform the CTE's of the individual plies to the laminate coordinate system.

$$\underline{\tilde{\alpha}}_k = \underline{\underline{T}}(\phi_k)\,\underline{\alpha}_k \tag{5.7}$$

Using Eq. 4.14 to obtain the transformed reduced stiffness and then using the ply thickness t_k, the effective CTE's of the laminate are determined according to:

$$\underline{\tilde{\alpha}}^{eff}(T) = \underline{\underline{A}}^{-1}(T)\sum_{1}^{n}\underline{\underline{\tilde{Q}}}_k(T)\cdot\underline{\tilde{\alpha}}_k(T)\cdot t_k \tag{5.8}$$

These effective CTE's of the laminate, $\underline{\tilde{\alpha}}^{eff}$, refer to the laminate coordinate system. As previously explained, the residual stress within each ply results from the mismatch of thermal strain between laminate and ply. Introducing the thermal strain of the laminate using the effective laminate CTE's to account for the total strain applied and the thermal strain of the individual ply into Eq. 5.5 leads to the residual stress state of the individual ply:

$$\underline{\sigma}_k^R = \int_{T^{\sigma 0}}^{T}\left(\underline{\underline{Q}}_k(T)\cdot\left[\underline{\alpha}^{sym:eff}(T)-\underline{\alpha}_k(T)\right]\right)dT \tag{5.9}$$

This equation refers to the ply coordinate system and hence $\underline{\tilde{\alpha}}^{eff}$ must be transformed using Eq. 5.7. It is important to note that the simplification of accounting for the stress-free temperature considers the effects of chemical shrinkage and viscoelasticity within the curing process in a simple form, as explained earlier. A more precise consideration of such contributions leads to more complex formulations, as reported in the literature (e.g. [10, 19]).

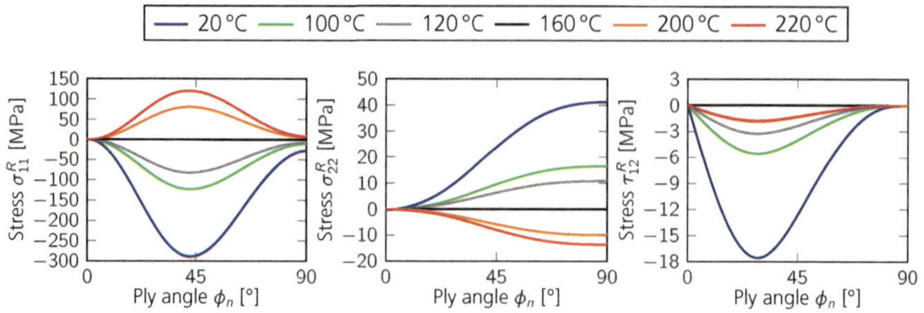

Fig. 5.7 Residual stress resulting from temperature change from $T^{\sigma 0}$ to T in dependence to the ply angle, exemplarily shown for the laminate $[0\,°_2 \mid \pm n]_S$

Equation 5.9 is useful for studying the residual stress development in MD laminates and its contribution to the overall load carrying capability of a material. Figure 5.7 shows the maximum residual stress that develops in a symmetric, balanced laminate at different temperatures. The orientation of the central plies (ϕ_n) is varied and displayed on the x-axis. A value of 0° represents a UD laminate, while a value of 90° represents a CP laminate. The residual stress shown in Fig. 5.7 do not include external mechanical constraints and result to the laminate internal restrictions, only. At temperature equal to $T^{\sigma 0}$, the residual stress is zero for all stress components, and over all ply angle variations. The UD setup does not show any stress development, as the plies are free to expand and no internal constraints are active. However, for all other variations, temperatures above or below the stress-free temperature cause residual stress to develop. In the fiber direction, the magnitude of residual stress increases for fiber angles up to 45° and decreases at higher orientations, resulting in a small value for the CP setup. This is due to the relatively small stiffness of the 90° layer. Cooling below the stress-free temperature results in compressive stress, while higher temperatures cause tensile stresses within the fiber direction. In the transverse direction, temperatures below the stress-free temperature cause tensile stress. This is a result to the relatively high CTE in transverse direction and restraint of the resulting contraction. The maximum stress is observed in the CP setup, where the interaction within the laminate stacking causes the highest mechanical restriction. Cooling to 100 °C can almost fully consume the transverse tensile strength capability in relation to the *A-value* allowable, and cooling to RT can even lead to stresses beyond. Compressive stress developed for temperatures above the stress-free temperature is less critical. The highest shear stress is observed at 30° ply angle and scales with the applied temperature difference.

Figure 5.8 illustrates the evolution of residual stresses over temperature for different laminates. These laminates are used previously to study their stiffness evolution over temperature, conf. in Fig. 4.8 and Table 4.4. These laminates represent the extreme conditions and the scaling of stress is consistent to Fig. 5.7. The UD laminate exhibits no residual stress as seen before. The MAP laminate and the CP laminate represent the laminate stackings causing the maximum stress conditions

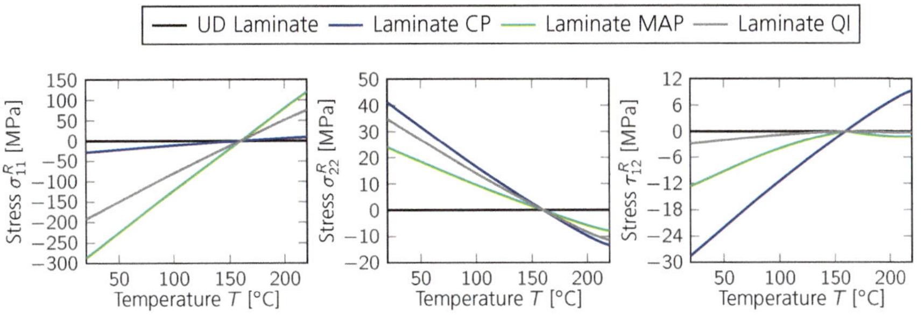

Fig. 5.8 Maximum residual stress resulting within exemplaric laminates through temperature changes from $T^{\sigma 0}$ to T

in fiber direction and transverse direction, respectively. The QI laminate is added for informative reason, as it is closer to realistic laminate stackings. Again, residual stress is zero at the stress-free temperature for all configurations. The direction of stress that occurs, i.e. tensile or compressive, is similarly consistent. The magnitude develops almost linear over temperature but depends slightly on the property relations over temperature. Notably, among the configurations considered, the QI setup displays the lowest levels of shear stresses.

5.2 Stability

The stability behavior is a critical aspect that must be taken into account when designing lightweight structures. Stability analysis determines the ability of structures to withstand specific load conditions without experiencing spontaneous deformation. When such structures become unstable, they undergo a transition from an in-plane deformation state, which experiences membrane stress, to an out-of-plane deformed state that involves superposed membrane and bending stresses [20, 21]. This can lead to critical stress, severe damage, or total collapse. Therefore, the stability behavior is a crucial design factor.

This section presents an overview on the stability problem of structures in thermomechanical load conditions. A review of literature related to stability analysis under thermomechanical conditions provides a coarse overview of the current state-of-the-art in the field. As buckling is an important design criterion, various methods for evaluating the buckling load in distributed thermal conditions are discussed. These methods form the basis for postprocessing and are used in the structural analysis to assess local buckling.

Problem Description Under Thermomechanical Conditions

The buckling bar is a well-known example of stability phenomenology [20] and it is useful for illustrating the problem resulting from thermomechanical conditions and related sensitivities. Figure 5.9a depicts a bar in its original, perfectly straight

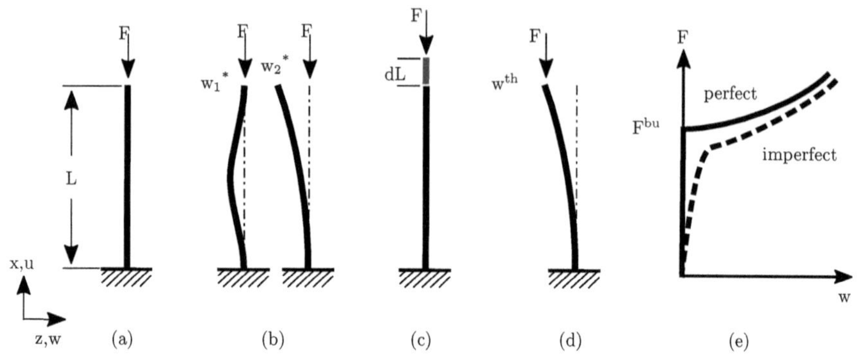

Fig. 5.9 Exemplaric cases of the buckling bar: (**a**) perfect shape, (**b**) imperfect shape, (**c**) perfect shape and uniform heating, (**d**) perfect shape and heating with thermal gradient T(z) (**e**) load-deflection curves highlighting the sensitivity of imperfection. Based on and modified from [20]

shape when subjected to axial compressive load. The bar remains in a stable state until it reaches a critical load, denoted as F^{bu}. At this load, as shown in (e), the state becomes indifferent, and further perturbation or load increase leads to a transition into another stable state, known as bifurcation. Typically, this state involves a wavy shape due to out-of-plane deformation and is commonly referred to as buckling. The load regimes leading up to and following buckling are referred to as the pre-buckling regime and post-buckling regime, respectively. Equation 5.10 represents the column buckling equation [20]. It is a mathematical expression that is commonly used to calculate the buckling load of the buckling bar and to highlight its sensitivities.

$$F^{bu} = n \cdot \frac{\pi^2 EI}{L^2} \tag{5.10}$$

In Eq. 5.10, E, I and L are the Young's modulus, the moment of inertia and the length of the bar. The parameter n accounts the interaction of BC on the buckling load.

Imperfections can have a significant impact on the buckling load, as illustrated in Fig. 5.9b by the geometrical shapes w_1^* and w_2^*, which deviate from the nominal structure. The influence of such imperfections can be observed in the load-deflection curve shown in (e), which results in lower structural stiffness and different bifurcation behavior.

In addition to mechanical loads, buckling can also occur due to uniform thermal loads, as shown in (c) of Fig. 5.9. When a structure is constrained, the thermal expansion results in mechanical strain and stress. In the case of a perfectly uniform thermal load and material properties that are independent of temperature, the resulting stress state is equivalent to that of a purely mechanical load, as shown in (a). To incorporate the effect of thermal loads, the thermal strain relation and Hooke's law can be introduced into Eq. 5.10. Rearranging yields:

$$\Delta T^{bu} = \frac{n^2\pi^2 I}{L^2 A\alpha} \tag{5.11}$$

with A and α as the cross section area and the coefficient of thermal expansion. This relation describes the critical temperature change that leads to buckling. This effect is commonly designated to as thermal buckling and typically refers to uniform thermal conditions.

However, the occurence of temperature in distributed shape is likely in real applications and a central aspect of this thesis. Liu et al. [22] developed analytical formulations to investigate the buckling behavior of a bar under thickness-wise temperature gradients. The resulting thermal moment and degradation of elasticity caused by temperature are observed to have significant sensitivities. Figure 5.9d illustrates the deformation wth of a bar that is clamped at one end and subjected to thickness-wise temperature gradients. The similarity between (d) and (b) suggests that these temperature distributions can act as imperfections, leading to deflections that deviate from the nominal shape, and therefore can be sensitive to the buckling behavior.

Literature Overview

There is a significant amount of literature available on the stability behavior of plates and shells under thermomechanical conditions. These stability problems are similar to the buckling bar but more complex due to the two-dimensional nature of the problem [20, 21]. The stability of these structures is dominated by properties in the loading direction, but it is also influenced by the transverse behavior, especially in the case of thermal loading as a consequence of thermal expansion. Chen et al. [23–25] investigate the thermal buckling and post-buckling behavior of symmetric and antisymmetric plates under uniform heating. Temperature distributions with in-plane gradients and their sensitivity on the effective buckling temperature of composite plates are investigated in [26]. Other investigations explore the applicability of strain superposition for combined loads of uniform heating and uniaxial compression [27–33]. Singha et al. [34] demonstrate the sensitivity of through-thickness temperature gradients on the buckling and deflection behavior. The temperature dependence of elastic properties on the buckling behavior is invesitgated within [35–38]. Sensitivities resulting to the use of different shear deformation theories are investigated in [33, 37–39]. They may become relevant for specific laminates and high temperatures. An extensive review of Noor and Burton [40] emphasizes the sensitivities mentioned above.

The orthotropic nature of the laminate and resulting effective, thermal expansion and stiffness behavior, material property degradation over temperature and resulting thermal stresses all affect the stability of plates and shells. Although only a few of these investigations specifically refer to distributed conditions, it becomes obvious, that there are in fact sensitivities arising from distributed conditions which may become relevant for very specific structural setups and load conditions. The simple enumeration of these sensitivities emphasizes the complexity that is created when considering temperature effects, as all of these factors interact with one another.

There is a wealth of research on the buckling behavior of thin-walled structures at RT conditions, particularly stiffened panels and cylindrical shells, with numerous experimental and theoretical studies investigating their buckling behavior at RT conditions, e.g. [5, 6, 21, 41–43]. However, few studies have considered the effect of thermomechanical conditions on the buckling behavior, and most investigations are based on simplifying assumptions such as uniform temperature or independence of material properties to temperature. Singer et al. [43] provide an extensive overview for the buckling behavior of structures under thermomechanical conditions from a theoretical perspective and different experimental considerations. Shen and Li [33] investigated the buckling load sensitivies under uniform temperatures for cylindrical cross-ply shells. Chang and Card [44] have compared different stiffening concepts of cylindrical shells for arbitrary temperature distributions over thickness. Rolfes et al. [45] have examined the sensitivity of buckling to thermal preload and thickness-wise thermal gradients for different laminate setups. Breivik et al. [46] have numerically and experimentally investigated the buckling behavior of stiffened composite panels under different combined thermomechanical conditions, observing small but measurable deviations from the expected behavior.

The literature reviewed suggests that the sensitivities observed in simple plates are also applicable to more complex structures, such as stiffened panels. However, the amount of research conducted for complex structures is limited. It is insufficiently investigated in detail how the stability behavior is affected by the behavior of the stiffeners under distributed thermal conditions. For instance, it is unclear how the temperature-based reduction in stiffening affects the stability behavior, or how the global thermal deformation, which is also attributed to the stiffeners, influences buckling shape and buckling loads. The DiA method outlined in this thesis may be useful for investigating these sensitivities in future research. However, the current work focuses on flat shells, which represent skin field elements as previously discussed.

Criteria to Evaluate the Buckling Load

Stability analysis presents a significant challenge in determining the buckling load, especially when using approximating analysis techniques such as FEA. Unlike exact theories or methods that, for instance, utilize eigenvalues, the nonlinear FEA used here relies on discrete physical parameters obtained through analysis. The sensitivity of temperature distribution can potentially result in variations in the buckling behavior, which can make it challenging to determine the buckling loads accurately. These buckling loads therefore need to be determined in dependence of the thermal parameter space, and the evaluation methods employed must be capable of capturing the resulting sensitivities. Inconsistent deformation behavior and buckling patterns, as well as varying progression of structural stiffness, require suitable criteria for selecting the appropriate buckling load. Similar problems arise when optimizing structural designs with varying topology and laminate setup [47, 48].

In the present work, constant geometrical parameters and laminate setups are employed in each thermal case analysis (explained in Chap. 6). Each case considers a specific temperature range and examines the structural behavior under varying temperature distributions. The evaluation methods must be capable of capturing whether the buckling behavior is sensitive to these temperature distributions, while also distinguishing between thermal deformation and buckling.

Kriegesmann [42] suggests several methods to evaluate the global buckling of stiffened panels, such as examining stringer deformations, observing a drop in the load-displacement curve, or assessing in-plane stiffness reduction. Freund [48] compares the displacements at specific locations against predefined margins to classify the buckling behavior as local buckling, local stringer buckling or global buckling. However, these techniques involve discrete values applied to identify the global buckling state, which may not account for buckling at varying thermal deformation states. Therefore, a more reliable and convenient procedure is needed. Wilckens [6] proposes an approach that determines the onset of local buckling for stiffened panels by analyzing changes in the structural stiffness characteristics due to buckling. The structural stiffness K is defined by the axial load applied to the loading edge F_{ax} to compress the structure by a specific displacement u_{ax}.

$$K = \frac{\delta F_{ax}}{\delta u_{ax}} \tag{5.12}$$

The change in structural stiffness between two states is typically observed as a continuous function that transitions from a higher to a lower magnitude of stiffness. An exemplary load-shortening curve is presented in Fig. 5.10, showing the relationship between axial force (F_{ax}) and axial displacement (u_{ax}) as black line, and the resulting structural stiffness in blue. Additionally, tangents of the load-shortening behavior (black dotted lines) indicate the change from the pre-buckling to the post-buckling domain. In the prebuckling domain, which represents a state of membrane stress, the structural stiffness initially remains almost constant. However, as the load increases and buckling occurs, the structure deforms and transitions to a second stable state. This state exhibits a decreasing stiffness at increasing loads due to continuously increasing bending deformations. The inflection point of the structural stiffness curve can be used to identify the central point of this transition. This point corresponds to the minimum value of the stiffness change (red curve) and is determined by analyzing the derivatives of the curve.

$$F_{ax}^{bu} \rightarrow min\left(\frac{\delta K}{u_{ax}}\right) \rightarrow \frac{\delta^2 K}{u_{ax}} = 0 \tag{5.13}$$

This approach is particularly useful as it enables accurate determination of the critical point of the transition. To ensure conservative estimates, Wilckens further proposes to use the minimum value of the second derivative of the structural stiffness (green line), which corresponds to the zero-crossing of the third derivative (as shown in Eq. 5.14). This point corresponds to the most significant rate of stiffness change [6].

$$F_{ax}^{bu} \rightarrow min\left(\frac{\delta^2 K}{\delta u_{ax}^2}\right) \rightarrow \frac{\delta^3 K}{\delta u_{ax}^3} = 0 \tag{5.14}$$

This second method offers a more conservative representation for assessing local buckling. In Fig. 5.10, the two black arrows indicate the buckling loads evaluated by

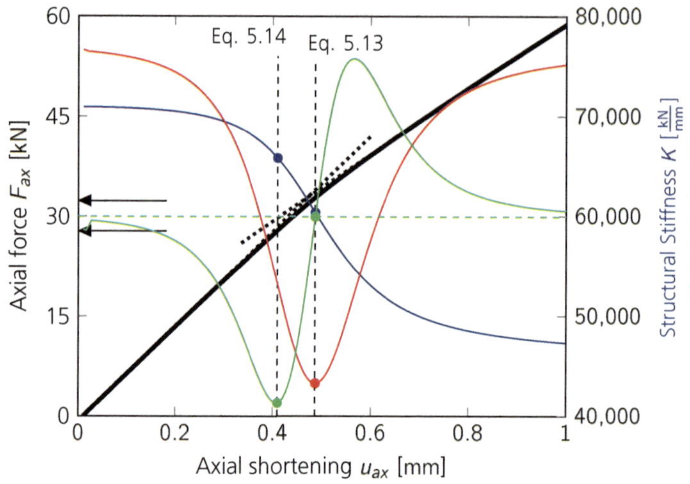

Fig. 5.10 Determination of the buckling point using the structural stiffness and its derivatives, according to Wilckens [6]

using the two approaches. Equation 5.13 leads to a buckling load above 30 kN while using Fig. 5.10 determines buckling at lower axial shortening and a corresponding buckling force below 30 kN. However, both are useful for assessing local buckling and they are suitable for situations where the buckling transition is influenced by varying thermal deformation states.

5.3 Material Failure

The material failure load decribes the limit state at which a material can no longer support a load due to strength failure. Predicting failure is crucial for efficient design, as it aims to utilize the load carrying capacity of materials efficiently but has to reliably ensure safety. Experimental failure assessment for MD laminates would be impractical due to the large number of tests required to optimize a laminate for overall structural design. For thermomechanical load conditions, the need to perform characterization at multiple temperatures would even increase the number of required tests. Therefore, failure prediction has to be achieved at a different scale, which also promises a better understanding of the phenomenology of failure. To this regard, UD laminates are commonly used for investigating the material properties of CFRP materials. The results are then used for instance through CLT or FEA tools to predict the behavior of MD laminates. However, testing at unidirectional load conditions allows for the characterization of typical strength properties of the UD lamina, including the inherent orthotropic behavior of the material and characteristic failure mechanisms. Such testing is conducted within the previously explained characterization program.

However, testing at multidirectional load conditions is complicated and elaborate and significantly increases the number of required tests. To address such load conditions, failure prediction methods are developed for practicality, utilizing the UD strength components to estimate composite failure at multidirectional loads.

In the case of MD laminates, it can be important to distinguish between initial failure, which marks the onset of any degradation, and final catastrophic failure, as highlighted by Soden et al. [49]. Due to the complexity and uncertainties involved in predicting damage progression of a laminate (the margin in between both), it is common practice to use the failure initiation of a single ply as a design criterion. This approach is often referred to as First Ply Failure (FPF) and is conservative, as the residual laminate can usually carry the released load at the moment of FPF. To simplify the problem, some assumptions such as linear elasticity are used, which are common and adequate approximations for brittle materials.

Multidirectional loads are common load conditions, and failure prediction must account for them. Many engineering approaches are available for estimating composite failure, providing highly diverse theories. The World-Wide Failure Exercise (WWFE) [49–57] contributes exceptionally to the development of composite failure prediction methods by providing an experimental database, by comparing and assessing different failure theories and by identifying limitations.[2] However, as indicated by Hinton et al. [53], there are research gaps, such as cracking under thermal loading and delamination, that need to be addressed. Temperature sensitivity is poorly addressed within literature due to the increased experimental complexity added by temperature.

This section aims to provide an overview of suitable failure prediction theories which can be used for estimating the FPF within structural analysis at thermomechanical load conditions. The developed material models are utilized to compare these theories at different temperatures. It is important to note that this study does not focus on enhancing these theories. However, the resulting insights can be used as a basis for future research on verifying and improving failure theories at elevated temperatures.

The high variety of existing approaches is based on a common mathematical foundation that describes the stress state as a multidimensional space. This space can typically be represented by a three-dimensional coordinate system reflecting plane stress conditions, $\underline{\sigma} = (\sigma_{11}, \sigma_{22}, \tau_{12})$, which is also the approach used in this work. To simplify the failure prediction process, it is assumed that the state of a ply within one local element is in a uniform thermal condition at a specified temperature and is therefore used for failure prediction. This is a reasonable assumption since structural analysis typically uses relatively detailed discretization. The actual stress state used for failure prediction $\underline{\sigma}^i$ is obtained by the stress vector between $\underline{\sigma}$ and the origin of the stress space, $\underline{\sigma}^0 = (0, 0, 0)$, which represents a stress-free state. The stress vector $\underline{\sigma}^i$ has a certain length and direction that describes the load state.

[2] Within the first WWFE [49–52] the scope was to investigate 2d failure theories for composite materials, the second WWFE aimed to focus on 3d failure theories [54–56], while the third WWFE [57] aimed to investigate matrix cracking criteria.

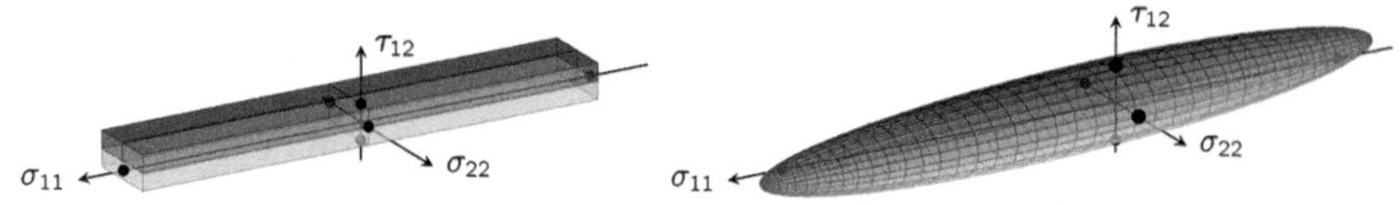

Fig. 5.11 Exemplaric failure envelopes. Points refer to the uniaxial strength components

The limit state of a material, i.e., its maximum load-carrying capacity, is defined by a closed surface within the stress space. A single point on this surface corresponds to a single stress state where failure just occurs and is given by $\underline{\sigma}^F$. Hence, the inner body of this surface represents the preserved material state, while the outer space represents the failed material. This is also known as the failure envelope or failure body. A rectangular failure body, created by six planes parallel to the coordinate axes, is the simplest example, while a more complex cigar-like failure body can also be obtained (conf. Fig. 5.11). Various failure approaches related to these failure bodies are be described in more detail later.

All failure approaches discussed in this work use the failure index f^i as a metric to distinguish between failed and preserved material. The failure index is obtained by dividing the stress vector $\underline{\sigma}^i$ by its corresponding failure stress vector $\underline{\sigma}^F$ given by identical directions. The different magnitudes representing the stress levels, allow to predict failure using the failure condition which is given by:

$$f^i = \frac{\underline{\sigma}^i}{\underline{\sigma}^F} \geq 1 \tag{5.15}$$

Equation 5.15 is applicabale to the full failure surface, i.e. using the entirety of possible stress vectors. Cuntze et al. [58] proposed the use of the reserve factor f^{res} to express the stress state of a material, which is the inverse of f^i and Eq. 5.15. The reserve factor represents a multiple of the failure stress, and it allows for more convenient understanding of the potential failure margins, as required in laminate optimization processes. However, within the present work, the failure index denotation of Eq. 5.15 is used in all cases.

Residual Stress Consideration

Residual stresses can have a significant impact on material failure [59] and Sect. 5.1.4 describes the origin of residual stress development on the meso-scale. To consider such residual stress in failure prediction, Puck suggests simple superposition of the constant residual stress and the actual load-dependent stress [60]. This effectively shifts the origin of the failure body by the magnitude of residual stress $\underline{\sigma}^R$ or, conversely, shifts the failure envelope by $-\underline{\sigma}^R$. Cuntze [61, 62] identifies two kinds of residual stress being sensitive to composite failure, both of which have been discussed at multiple points earlier in this work. Residual stresses of the first kind arise from curing and cool-down through the mechanical constraints within a MD laminate. To account for them, Cuntze assigns the cure temperature as a stress-free condition, i.e. by $T^{cure} = T^{\sigma 0}$, and applies superposition as proposed by Puck [60]. A similar, thermoelastic approach, utilizing $T^{\sigma 0}$ for residual stress consideration, is used here, conf. Sect. 5.1.4.

Residual stresses of the second kind, i.e. stresses between fiber and matrix, are not considered by Cuntze [61] since they could not be assessed. However, these stresses are expected to be sensitive to failure, as previously mentioned, and should be incorporated into the failure theory. All failure theories discussed here homogenize fiber and matrix and therefore cannot account for these second-kind residual stresses. Enhancing failure theories to include second-kind residual stresses could be a future endeavor to improve the accuracy of failure prediction in dependence of temperature.

5.3.1 Failure Theories for the Unidirectional Ply

This section provides an overview of selected failure criteria for predicting the failure of UD laminae. While the assessment of their predictive quality is not possible due to the lack of experimental data for multidirectional load conditions, these criteria have been chosen based on their applicability and prominence. The aim of this section is to give an overview and discuss the challenges that may arise at elevated temperatures and the related property degradation mechanisms.

Several comprehensive overviews of prediction theories for UD laminae failure can be found in the literature. One of the most extensive sources is the work around the WWFE, which describes more than 15 theories and related references (see footnote 2). Another good overview, is given by Cuntze et al. [58], which covers different criteria and experimental techniques for parameter determination. A condensed overview of the five theories discussed here can be found in Mittelstedt and Becker [7]. These theories coincide with the mostly recommended approaches of the WWFE. The following section describes the main aspects of these theories, while their mathematical formulations are provided in the Appendix Sect. B.

The simplest type of failure criterion is the Maximum Stress Criterion (MSC), which describes a rectangular fracture surface (see Fig. 5.11, left). It is defined through the uniaxial strength values and is given by planes parallel to the coordinate system (conf. Eqs. B.1 to B.3). As a result, the criterion does not account for interactions between individual stress components, which can occur towards the corners or edges of the stress space, while the failure mode can be distinguished based on the load direction. According to Zinoviev et al. [63], the simplicity of the approach is a positive aspect. This argument is supported by the complexity that comes with the orthotropic behavior of UD CFRP materials and the resulting shape of the failure surface which make it possible to capture a wide area of the failure envelope with the MSC. However, the criterion neglects the stress interactions that occur towards the corners or edges of the stress space. Therefore, in cases of specific multidirectional loads, such as at the corner areas of the failure body, the MSC might lead to distinct overestimation. The *Maximum Strain Criterion* is an alternative to the MSC and defines the failure conditions based on ultimate strains instead of failure stress components. This criterion takes into account the effects resulting from lateral contraction.

When engineers first developed composite failure prediction methods, it was crucial to have a simple yet effective mathematical approach. Tsai and Wu's criterion [64, 65] became widely known for using a tensorial expression of strength components to establish a single equation that describes the failure surface covering the full stress space. This approach is adaptable to different material laws, such as plane-stress or three-dimensional, and allows for the definition of interaction terms between individual stress components. The general formulation for the failure condition is given by:

$$F_i\sigma_i + F_{ij}\sigma_i\sigma_j = 1 \qquad \text{with} \qquad i, j = 1, 2, ..., 6 \tag{5.16}$$

where F_i and F_{ij} are strength tensors and σ_i and σ_j are the components of the stress vector $\underline{\sigma}$. The F-terms are not necessarily equal to the reciprocal of the uniaxial strength components but are typically based on experimental data to fit the approach, as for instance shown by Tsai and Wu [64]. The approach does not obtain information about the failure mode and therefore is only applicable to FPF estimation and can not be utilized further for progressive damage analysis. However, while the Tsai-Wu criterion is frequently used due to its simplicity and ability to consider stress interactions, it is important to note that the individual terms are often used without proper verification or questioning. For instance, some versions of the criterion consider interactions between shear and normal strengths, while others do not. Some of the interaction terms are also questionable from a physical perspective [66], which could lead to incorrect failure predictions. Therefore, empirical studies are needed to verify or adapt the interaction terms, especially for different thermal conditions, as the approach does not account for interactions between different failure modes. Nonetheless, the Tsai-Wu criterion is a step up from the MSC criterion as it at least considers the interaction of stress components.

Hashin [67] understood the importance of distinguishing between different failure modes, which was based on the observation of an oblique fracture plane occurring parallel to the fibers as a matrix failure mode. This observation can be compared to the Mohr-Coulomb hypotheses in brittle isotropic materials, where varying fracture plane angles θ_{FP} result in different stress component combinations [7, 68]. Hashin's failure criterion therefore assumes invariant failure behavior for stresses in the fiber and transverse direction, differentiating between fiber failure modes and matrix failure. In addition, the criterion considers tensile and compressive loading, leading to four Eqs. (B.5) to (B.8) that are representative of load directions in normal stress space. The four related failure modes address normal stresses, and a fifth mode would be obtained by pure shear. Interactions between normal stresses and shear stress are considered in Hashin's approach through quadratic approximation. The neglect of shear interaction on the fiber failure mode under compressive loads is due to a lack of data and knowledge at the time of Hashins study [67]. Further research, as descibed below, closes this gap by matrix compression failure mode approximations through more complex formulations. Hashin also indicates that higher-order approximations could produce better results, but the quadratic version presented here is mostly preferred due to its convenience and the necessary test effort required to determine the resulting higher order coefficients.

Puck's work is motivated by Hashin's observations, which similarly differentiate between Fiber Failure (FF) and matrix failure [60, 66]. Within the scope of Puck's *action plane concept*, the theory of matrix failure is further differentiated into related failure modes designated as Inter-Fiber Failure (IFF). Building upon the Mohr-Coulomb approach, Puck assumes that IFF always occurs within an inclined plane parallel to the fibers, which leads to the consideration of fracture plane inclination angles θ_{FP}. The conclusion, that failure is always caused within the plane of most severe stress conditions leads to a certain optimization problem related to θ_{FP}. Puck's differentiation between tensile and compressive IFF modes is similar to Hashin's theory. Tensile matrix failure is expected to occur at a zero fracture plane inclination angle, while for transverse compressive stress, a certain degree of shear interaction is addressed. This differentiation leads to the consideration of compression-dominated and shear-dominated failure modes. For fiber failure, Puck proposes a strain-based formulation, similar to the *Maximum Strain Criterion*, to consider at least lateral stress interactions. Additionally, he introduces empirically derived weakening factors to allow for material-specific adaptations by fitting to experimental data. There are different versions of Puck's theories available, and for more insights, the reader can refer to [60, 66, 69, 70]. These publications are valuable since they also address experimental verification as well as model parameter determination and their discussion. They also explore the possibility of fitting each criterion by adding these additional terms. For example, in [60], Puck adds an empirical shear term to consider the sensitivity of shear stress on the compressive strength causing kinking. Similarly, an empirical term is added to the IFF modes for considering longitudinal stress interaction (Fig. 5.12).

Cuntze aimed to develop a failure criterion that is simple and suitable for engineering design studies. His fracture-type strength criteria (FTC) proposes that each failure type should be addressed through an individual term, similar to Hashin and Puck's approach [71]. Cuntze identified five dominant failure types based on the dominant loading conditions: longitudinal and transverse failure under tensile and compressive loads, respectively, and shear loads. Later, he developed the *Failure mode concept* [61, 62, 72, 73], which contains an individual condition for each of the mentioned modes. Each individual failure condition is based on the formulation of invariants and is governed by a basic strength of the UD laminae [62]. The resulting terms for each failure condition are similar or partially equal to the other approaches. The criterion leads to the advantage that laminates can be optimized to individual failure conditions, since the resulting individual reserve factors obtain the design driving conditions. Cuntze's *Rounding-off* approach considers interaction between the modes and scales each term by an exponent. The superposition of all failure condition terms leads to a single equation and hence a closed failure surface, as shown in Eq. B.21.

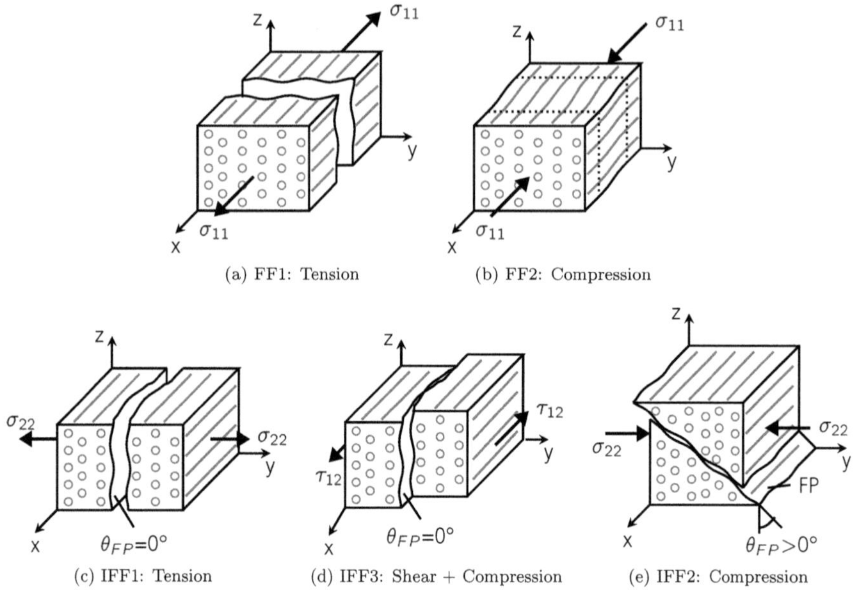

Fig. 5.12 Failure modes as distinguished by Puck [60] and Cuntze [72]. Fiber Failure (FF) is shown in the top row and interfiber fracture (IFF) in the bottow row. Arrows indicate the dominating stress component. The sketch is drawn in adapted form based on the version of Cuntze [62]

5.3.2 Discussion and Comparison in Dependence of Temperature

The orthotropic behavior and heterogeneity of composites make it necessary to distinguish between different failure modes for predicting their failure. The different fracture types of the UD lamina are governed by the dominant behavior of the fibers, the matrix, or their interfacial behavior. This differentiation is especially important at elevated temperatures due to the significantly different temperature dependence of fiber and matrix properties. Except for the Tsai-Wu criterion, which does not distinguish between different failure modes, all of the theories discussed above classify different failure modes based on specific load conditions (dominant failure type). Among these theories, the simplest criterion is the MSC, which uses only the UD strength components to predict failure and leads to five failure modes. On the other hand, the criteria developed by Hashin, Puck, and Cuntze distinguish between FF and IFF modes, which seems reasonable, especially when referring to the temperature-dependent material behavior.

A general agreement among all criteria is that FF is dominated by the related unidirectional strengths (σ_{11t} and σ_{11c}) while being not or only minorly sensitive to transverse or shear stresses. Figure 5.13 illustrates the sensitivity of the failure envelopes to interactions between stress in the fiber direction and shear stress, e.g.

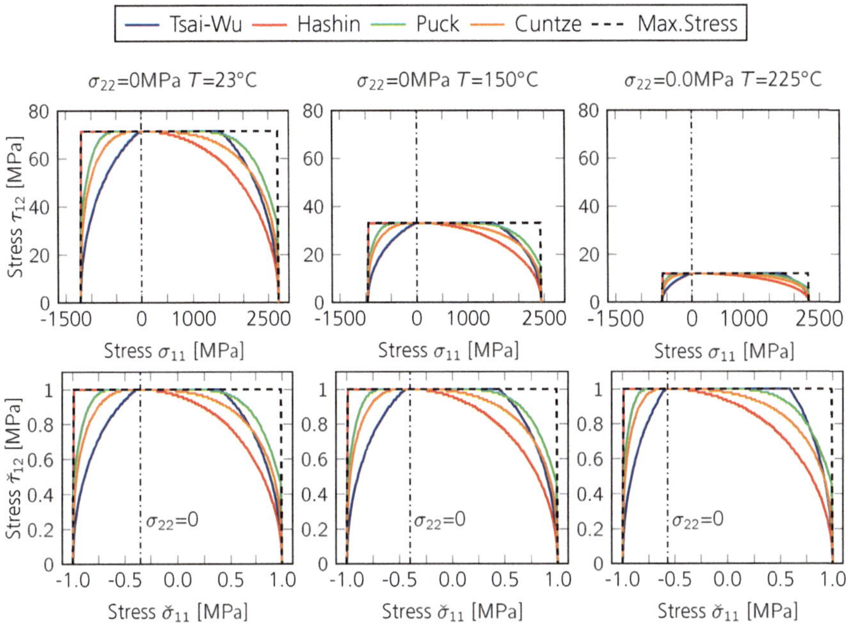

Fig. 5.13 Comparison of the failure prediction regarding typical FF modes in dependence of temperature. The top row presents absolute magnitudes and the bottom row shows normalized values

within the stress space of $\underline{\sigma} = (\sigma_{11}, \sigma_{22} = 0, \mid \tau_{12} \mid)$. In addition to showing the absolute values that indicate the degradation of the failure load over temperature, normalized values are also used to better highlight deviations of the trends over temperature.

The MSC formulation creates the outermost envelope, clearly stating no interaction between longitudinal loads and shear stress. The experimental results obtained by Michaeli et al. [74] partially support such behavior, but they do not include high shear stress combinations for compressive stress states. Hashin understood that simultaneous shear stress reduces the tensile strength, while using the MSC approach for compressive stress (due to missing data). Although Puck's tensile FF theory is equal to the MSC and does not contain a shear term, a certain shear dependency is noticed. This part of the Puck's failure envelope describes the IFF Mode A (conf. Eq. B.12), resulting from the use of the weakening factor term that addresses the longitudinal stress. Puck further emphasized in [60] that the compressive strength is similarly influenced by shear terms. Earlier, Edge [75] reported similar observations and a switching failure mode from pure compressive splitting towards a kinking mode occurring above a certain level of simultaneous shear stress. The explanation for this phenomenon is similar to the tensile behavior, although the resulting effect is different. A loss of fiber support resulting from local interfacial shear failure causes a macroscopic material failure. Thus, elevated shear stress can cause lower longitudinal strength. The sensitivity of tensile strength to interfacial stress is indicated by Gibson referring to its temperature dependency [76] (conf. Sect. 3.5). For compression, higher shear loads and consequently more interfacial failure cause fibers to kink instead of observing explosive fiber failure modes. This observation also explains the rapidly decreasing compression strength found around PES T_g occurring in combination with delamination and kinking. Introducing a certain dependency on shear is therefore reasonable. Both Hashin and Puck assume the FF modes to occur invariant of transverse stress. Compared to them, Cuntze similarly uses the MSC formulation but considers a certain conservatism regarding shear stress and transverse stress interactions using the rounding-off approach.

Referring to Fig. 5.13, the failure envelopes of Tsai-Wu, Hashin, Puck, and Cuntze follow a similar trend with two exceptions. First, as mentioned earlier, Hashin's criterion exhibits a deviation within the compressive stress domain following the MSC. Second, the Tsai-Wu criterion shows shear dominated failure within the tensile stress space until a certain tensile load margin is reached and interaction is introduced. This results in a stress range at 225 °C where the criterion describes the outermost curve and represents the least conservative formulation. This trend is less severe at lower temperature Puck's and Cuntze's theories follow a similar trend, and the differences can be attributed to the selected model parameters. Hashin's criterion shows the most conservative trend within the tensile stress domain, while the Tsai-Wu criterion has the lowest curve progression within the compressive stress domain and the most conservative criterion at these loads.

Figure 5.14 illustrates the behavior of IFF within the stress space perpendicular to the fibers, i.e., in the transverse and shear directions, represented by $\underline{\sigma} = (0, \sigma_{22}, \mid \tau_{12} \mid)$. Hashin's approach also distinguishes a dominant shear failure at compressive loads, although it does not have a specific equation as Puck's approach does. At low temperatures, Hashin's criterion is the most conservative for interactions between compressive and shear stresses. However, as the temperature increases, the part of the curve describing such interactions (values inside the MSC boundaries), shifts towards those boundaries. The failure envelopes of Tsai-Wu, Puck, and Cuntze criteria are similar across the entire domain and all temperatures, and they progress relatively close to the MSC, indicating a low interaction of the individual load types with the failure modes. This observation means a relatively low interaction sensitivity of shear stress and transverse stress on the failure behavior and supports the differentiation of three IFF modes, representing the left, upper, and right borders of the shown domain. In contrast, for FF, as shown in Fig. 5.13, the curve shape is more stable, indicating the presence of two related FF modes.

The failure initiation in composite materials is a complex phenomenon that involves different failure mechanisms on a microscale, including matrix cracking, fiber-matrix debonding, and fiber fracture. Since the initiation of failure is very local, it cannot be directly measured during characterization, and therefore, it is defined as the point where the initial failure propagates abruptly, causing macroscopic damage or failure. The criteria mentioned in the literature are specifically designed to address this point, particularly for the UD lamina. Applying the FPF criterion is the most

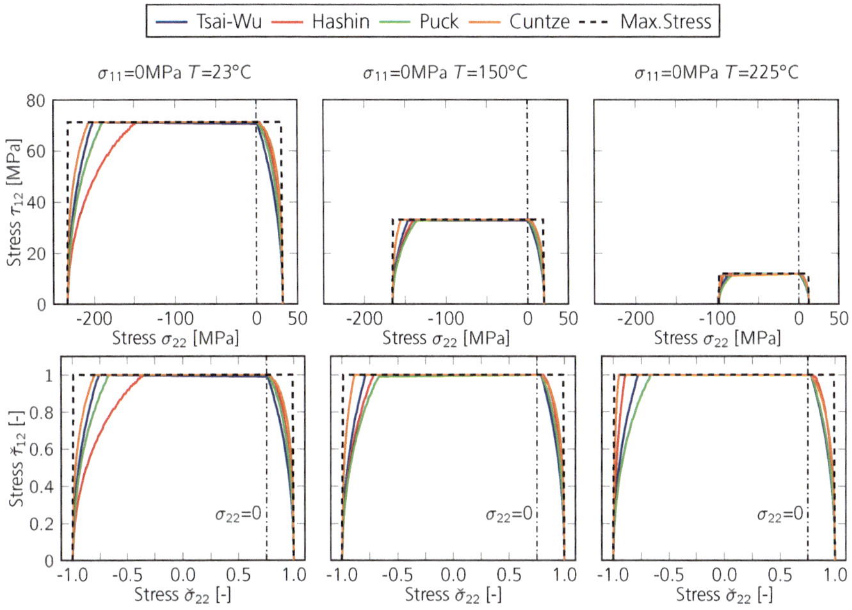

Fig. 5.14 Comparison of the failure prediction regarding typical IFF modes in dependence of temperature. The top row presents absolute magnitudes and the bottom row shows normalized values

conservative approach for their utilization to predict failure initiation in a MD laminate. It is important to note that neighboring plies in a laminate support each other, and loads are partially redistributed, which affects the sensitivity of the IFF failure modes to the laminate setup [77]. It is also worth noting that the FF of a single ply mostly leads to overall laminate failure, according to both Puck and Cuntze [62]. Therefore, differentiating between IFF and FF as design criteria is essential. For sizing procedures, Soden et al. [49] recommend using the criterion with the innermost fracture surface for being conservative.

The criteria discussed assume a perfect fiber-matrix connection and estimate failure initiation defined in equity with macroscopic failure. The statistical character of failure initiation and the propagating character within brittle materials allow such formulation. However, since many parameters influence the behavior, partially empirical factors are introduced to improve the accuracy of the predictions and to account for the inherent peculiarities of different material systems. The mentioned literature provides multiple verifications of this procedure for different criteria, materials, and also under combined static stress load cases.

The dependence of temperature on the failure criteria of plastic materials is a less explored topic within literature. Depending on the material, matrix materials tend to exhibit more ductile behavior at elevated temperatures, which may render the criteria unsuitable in such cases. Some publications [60, 73] suggest the use of nonlinear stress-strain curves to account for this behavior, but this is not a well-established process. Moreover, the range of applicable material behavior in terms of brittleness or ductility has not been defined or investigated. The consideration of kinking at compressive loads in fiber direction, which was observed during the present test program, could further improve the overall failure prediction accuracy and some related investigations are found in [76, 78–83]. Investigating these factors could help gain confidence in the application of these criteria at elevated temperatures, and is therefore an area for future research.

5.4 Probabilistic Analysis

The current sizing approach for aircraft components uses a *worst-case* thermal scenario within deterministic analysis which is represented by a uniform distribution of maximum service temperature. However, this *worst-case* scenario is only a contingency that may not even occur during the lifespan of the aircraft. Nevertheless, it must be considered in the design process due to its potential impact. An alternative approach, discussed in this work, replaces the deterministic design based on uniform temperature with a probabilistic approach that takes into account spatially distributed temperature variations. Such variations are more realistic but also more volatile, as they depend on factors such as ambient conditions and thermal boundary and load conditions. According to the laws of physics, a local heat source naturally leads to a hotspot with a maximum local temperature. The temperature shape of the hotspot region, i.e., the temperature distribution, can be described by several parameters as for instance maximum temperature, average temperature, thermal gradients in any direction etc. However, determining the most critical thermal state for the structural design and its parameters is initially unclear, and therefore, a large set of possible variations must be considered to ensure safety. To address this, a systematic approach can be adopted, which involves conducting a set of analyses that cover a number of possible variations of the input space. This can be achieved through a *design of experiments* methodology that creates samples of thermal loads and enables investigating the relationship between the thermal state and design values. These studies involve deterministic FEA for each sample, that applies distributed thermal loads, as previously explained.[3] The findings from this study can serve as valuable inputs for conducting sensitivity analyses aimed at identifying the most critical scenarios. Additionally, the results form the basis for carrying out an efficient probabilistic analysis that generate statistical design limits, i.e. allowable values of the design parameters, unlike the traditional deterministic approach. The methodology employed to derive such design allowables relies on the construction of a surrogate model, and a comprehensive description of this approach is outlined below. Similar procedures

[3] The thermomechanical FEA procedure is explained within Sect. 5.1. The procedure to apply thermal load conditions is described in paragraph Sect. 5.1.1 and the methods used to derive the design values are discussed in Sects. 5.2 and 5.3.

are commonly used in the literature, with the primary motivation being to challenge deterministic design approaches. One such example is to consider the effect of defects introduced through automated fiber placement [84].

Procedure

Each sample obtained by the design of experiments conducts a FEA and leads to a pair of input parameters X and resulting design values Y. Collectively, the dataset generated from a series of such analysis serves as the basis for the surrogate modelling approach and forms the training points.

$$(\underline{X}, \underline{Y})^{tp} \tag{5.17}$$

The training points must reflect the temperature dependency of the mechanical structural design values. To achieve this, the input parameters are chosen based on the parameters that describe the shape of the resulting temperature distribution. This formulation reduces the number of required parameters from all the dependencies needed to define the thermal boundary conditions and load conditions to a smaller number that is sufficient to represent the temperature characteristics. Each point in $\underline{X}^{tp}$ is represented by:

$$X^{th} = \{X_1, ..., X_n\} \tag{5.18}$$

Here, $X th$ represents a single parameter of the resulting temperature distribution and X_i defines an individual value obtained from analysis. After conducting the mechanical analysis, the structural behavior is determined, and postprocessing is performed to obtain the mechanical design values. These parameters include the forces of local buckling onset (Local Buckling (LB)), matrix failure (inter-fiber fracture, IFF), and fiber failure (FF). They are summarized as follows:

$$Y = \{F_{ax}^{LB}, F_{ax}^{IFF}, F_{ax}^{FF}\} \tag{5.19}$$

with Y^j indicating one of the above mentioned design values which are investigated throughout this work. Each design value stored within Y is individually linked to the input parameter set because they may have a different temperature dependency. As a result, different training points are formulated for each design value, and they serve as the basis for three separate surrogate models. Consequently, conducting n analyses results in three sets of training points.

$$\left(\{X_1^{th}, ..., X_n^{th}\}, \{F_{ax,1}^{LB}, ..., F_{ax,n}^{LB}\}\right) \tag{5.20}$$

$$\left(\{X_1^{th}, ..., X_n^{th}\}, \{F_{ax,1}^{IFF}, ..., F_{ax,n}^{IFF}\}\right) \tag{5.21}$$

$$\left(\{X_1^{th}, ..., X_n^{th}\}, \{F_{ax,1}^{FF}, ..., F_{ax,n}^{FF}\}\right) \tag{5.22}$$

The dataset given by the discrete training points is used to construct a surrogate model that expresses the functional relationship between inputs and outputs. In this study, the surrogate models capture the sensitivity of spatial temperature behavior on design values:

$$\hat{Y}^j = f\left(X^{th}\right) \tag{5.23}$$

It is important to note that surrogate modelling assumes the training points to be exact, even though their precision is limited by the accuracy of the FEA. However, because the surrogate model expresses a functional relationship between design values and within the space of temperature distribution, it is useful for efficient probabilistic analysis. By expressing the input parameters with a suitable probability density function, the resulting statistical distribution of a design value can be written as: It is important to note that surrogate modeling assumes the training points to be exact, even though their precision is limited by the accuracy of the FEA. However, because the surrogate model expresses a functional relationship between design values, it is useful for efficient probabilistic analysis. By expressing the input parameters with a suitable probability density function, the resulting statistical distribution of a design value can be written as:

$$\Phi\left(\hat{Y}^j\right) = f\left(\{\Phi\left(X_1\right), ..., \Phi\left(X_n\right)\}\right) \tag{5.24}$$

The *Monte Carlo Method* is the most straightforward approach for analyzing the statistical distribution of design values. In this method, Eq. 5.24 is repeatedly solved using random values sampled from the PDFs of each input parameter. Because this is performed using a surrogate, i.e., an analytical function, a large number of iterations, even up to millions or more, can be completed in a matter of seconds. This highlights the benefits of utilizing surrogate modeling techniques over traditional deterministic FEA, where individual analyses may take minutes or even hours, making a probabilistic analysis impractical.

After conducting a Monte Carlo analysis, a dataset $\underline{\hat{Y}}^j$ is obtained, representing the statistical distribution of a design value in terms of discrete values. To utilize this data in the design process, it must be converted into design allowables. These allowables are defined using quantiles, where the resulting allowable represents the margin that defines the limit state for a specified number of values obtained from the statistical representation. For example, when defining material property allowables using *A-value* or *B-value* definitions, 99 % or 90 % of all values, respectively, are higher or equal to the specified allowable. Using the quantile q, design allowables can be determined according to:

$$Y^{j,Alw} = \int_{1-q}^{\infty} \underline{\hat{Y}}^j d\hat{Y}^j \tag{5.25}$$

The resulting allowable value considers the variation in the physical behavior of the system obtained through the surrogate model and the input parameters. As the material parameters are applied using equal margins of conservativity, similar to the deterministic reference analysis approach, they can be directly compared to common design values. The designation of the design value as PDF, $\Phi\left(\hat{Y}^j\right)$, introduced by Eq. 5.24 is used to highlight the dependency between the statistical distribution of the input parameters and the resulting design value distribution. However, the fitting of a PDF to the discrete design value data $\underline{\hat{Y}}^j$ is not conducted and the allowable design values $Y^{j,Alw}$ are determined using the data as expressed within Eq. 5.25.

Surrogate Modelling and Accuracy Evaluation

The approach used for material modeling, as discussed in Chap. 4, is similar to the surrogate modelling technique described here and could also be represented using Eq. 5.23. However, since the temperature dependent property is a one-dimensional problem, it only highlights some of the challenges compared to surrogate modelling for more complex relations as for structural design values depending on temperature distributions. In the case of material modelling, the input parameter is temperature ($X \hat{=} T$), while the property being modelled represents the design parameter ($Y \hat{=} P$) and is determined at multiple temperatures to generate the training data. In addition, the material behavior provides a clear mathematical formulation for the training (fitting) procedure governed by the physical behavior of the UD material. However, the surrogate modelling procedure for structural design values differs when considering the temperature distribution due to the higher dimensionality of the related input space Xth. This higher dimensionality affects the number of required training points and computational effort needed to perform surrogate modelling with suitable accuracy and reliability. Compared to the material model analogy, the design value dependency of thermal input parameters is unknown and dependent on factors of the structural system, such as orthotropic material properties, selected laminate setup, and applied boundary conditions. Therefore, it is essential to have suitable metrics and procedures to check the accuracy of the surrogate model. Additionally, the source of the data used for the surrogate model, FEA, which is assumed to be exact, introduces different error sources compared to experimental measurements. Even though assuming suitable accuracy of the design criteria, a systematic error is obtained due to the approximative nature of the incremental analysis procedure. Therefore, it is important to explain the basic principles used to assess the accuracy before addressing the surrogate modelling process in greater detail.

The mentioned approximative nature of FEA means that the quantities used to assess the design values and representing the structural behavior are expressed at discrete points (increments) over the domain of load and boundary conditions. The accuracy of the resulting prediction, such as the entrance of a criterion, depends on the step width of the increments. Although the exact physical solution would likely lie between two increments, no additional interpolation technique is considered, resulting in a bias in the training points. This limitation restricts the ability of surrogate modelling to accurately represent the physical relation. For instance, local buckling is detected through FEA postprocessing at load $F_{ax} = bu^{LB}$, which is related to the increment r, and the criterion correlates this to buckling failure. Therefore, the true onset of buckling would lie between increment r and its prior increment $s = r - 1$ where buckling does not yet occur. This systematic problem exists in similarity for strength failure design values. Consequently, the uncertainty of a design value Y^j arising from the incremental model procedure, ψ^j, is governed by the incremental difference of the evaluated quantity (here, axial force F_{ax}):

$$\psi^j = Y_r^j - Y_s^j \qquad \text{with} \qquad Y_s^j, Y_r^j \in F_{ax}^j \tag{5.26}$$

This value is obtained from the incremental model procedure, representing the analysis uncertainty ψ^j of a design value Y^j, but it is not similar to the material uncertainties captured earlier through experimental characterization. It rather expresses the maximum margin of variance that may potential be present for the assessed design value. It is essential to recognize this distinction because the precision of design value surrogate models is limited by the quality of their training points. Further, this understanding enables the definition of error margins that are useful for rating the accuracy of the surrogate models. A model with high precision can be defined as having a mismatch between the model and training data that is lower than the model uncertainty.

$$\hat{e}^j = f\left(\hat{Y}^j, Y^j\right) < \psi^j \tag{5.27}$$

Model errors can be quantified using different metrics and strategies. There are two main objectives behind error assessment. First, the accuracy of the surrogate model must be determined to feedback the derived allowables by a measure that quantifies its accuracy and reliability. Second, the number of training points and related FEA scales the cost of conducting a design iteration. Predefining a minimum required accuracy and selecting tailored training points can lead to more efficient approaches with a lower number of training points. In this regard, Freund [48] proposed an iterative and adaptive training point definition. The first objective is a major concern in the present work, but it is closely linked to the second, since error assessment can directly be exploited in schemes addressed through the second point. Therefore, a suitable error assessment is essential for both objectives. However, the second point serves only for information and is not further investigated here.

Two basic procedures to determine errors are used in this work. The first is the Leave-One-Out (LOO) strategy, also known as cross correlation [48, 85], in which a specified number of training points (one or more) are omitted for model training and are later used for model error quantification. In the present work, a single training point is used and the procedure is repeated for every training point of the input dataset. For the Kriging models used here, this procedure does not determine a true error but tests the sensitivity of the surrogate model against changes. It ensures the model accuracy between the training points to be at least equally precise as the determined error. A high local error, for instance, can be caused at spaces with comparably low number of training points and high sensitivity of the parameters on the design value. The second procedure used for error assessment is the use of additional test datasets $(\underline{X}, \underline{Y})^{test}$ which are defined in reference to Eq. 5.17. An arbitrary number of FEA differing from the training points is used purely to assess the surrogate model accuracy while they are not part of the surrogate models training dataset. The resulting errors refer directly to the trained surrogate model but are only as representative as their distribution within the uncovered parameter space.

To evaluate the individual errors, i.e., to determine the magnitudes of $\hat{e}^j$, a combination of local errors and averaged global errors is utilized providing a better understanding of the error distribution within the parameter space. However, unlike the material model assessment (Sect. 4.2.4), absolute error metrics are employed here. Absolute error metrics are useful in this context because they facilitate the comparison of the model against the analysis uncertainty ψ^j, by incorporating errors into Eq. 5.27. The Maximum absolute error (MaxAE) measures the maximum local deviation within the input space, while the RMSE calculates the model error averaged over the entire input space.

$$\hat{e}^{MaxAE} = max\left(\mid \hat{Y}^j - Y^j \mid \right) \tag{5.28}$$

$$\hat{e}^{RMSE} = \sqrt{\frac{1}{n}\sum_{i}^{n}\left(\hat{Y}^j - Y^j\right)^2} \tag{5.29}$$

The complexity of the structural system behavior and the multidimensionality of the problem make model selection and model training challenging. Forrester et al. [85] provides an extensive overview of different surrogate modelling techniques, while Freund [48] and Liebisch [47] used Kriging techniques to investigate the stability behavior of stiffened shell structures over a wide range of geometrical and material parameters. Kriging initially applies a regression approach fitted to the dataset. The difference between the resulting regression model and the dataset is assumed as a statistical error [48], which is compensated by a superposing term done for each training point. The Kriging models are exact at the training points, with statistical terms formulated as weight functions that fade with distance to the training point to cover the space between the points. A correlation matrix is used to consider a certain superposition between the different training point functions. Hyper-parameters are optimized to fit the model. Figure 5.15, visualizes the approach. A target test function is given by five training points. The Kriging model is exact in its training points and the error increases in distance to the training points. A more detailed explanation can be found in [48]. For the selected FEA procedure, the investigated structure, the related input paramters and the considered design values, a simple Kriging methodology with linear regression functions as the surrogate type of choice is used here. This choice is made due to the relatively intuitive fitting procedure for unknown system behavior and its good accuracy within the parameter space (extrapolation should be avoided). Further details on model precision are discussed in the results section of Chap. 6. However, it has to be emphasized, that the surrogate model accuracy highly depends on the database and the complexity of the problem, e.g. the number of parameters defined, the number of training points available and the shape of the response function that has to be modelled. The amount of available data, i.e. FEA, is limited here. It is

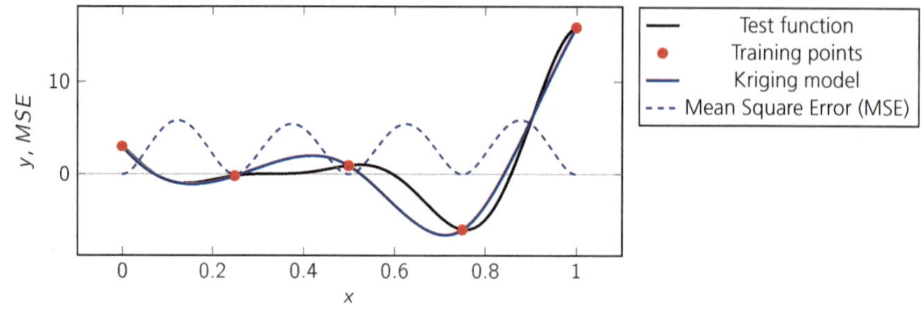

Fig. 5.15 Visualization of the Kriging surrogate modelling approach used. According to Freund [48]

therefore essential to formulate the requirement of consistent deformation behavior for the application of surrogate modeling. This condition limits the complexity of the response function by avoiding discontinuities or critical nonlinearities within the response function.

5.5 Analysis Procedure to Derive Structural Allowables

The previous sections outline the individual analysis steps and evaluation procedures developed throughout this work, which are necessary for a comprehensive assessment of structural behavior under distributed temperature conditions. Combining these steps and procedures finally enables the related evaluation of structural behavior, and allows to derive structural allowables that are representative for such conditions. The purpose of this section is to provide a concise overview of the overall analysis procedure, connecting the individual tasks and methods into a unified framework.

Figure 5.16 provides a visual representation of the overall framework, which includes the individual analysis procedures and the colors used align with the overall overview explained in Sect. 2 (conf. Fig. 2.6). The workflow begins with a definition task (Input definitions, red box), where the structural part, thermal and mechanical load domains, and boundary conditions are defined. Additionally, knowledge of the material behavior is necessary, which requires determining related temperature-dependent material models, as explained in Chap. 4 and highlighted here by the green box. This material behavior serves as input for deterministic FEA, yellow box, which includes a postprocessing step to derive the corresponding design values. This step involves conducting a specified number of repetitions to produce a set of training points, with each analysis considering a different temperature distribution and leading to related design values. Based on these training points, surrogate modelling is performed to generate a function (surrogate model) for each design value over the thermal parameter space. It is important to conduct error assessment to ensure the accuracy of the surrogate models. If necessary, additional analysis can be conducted

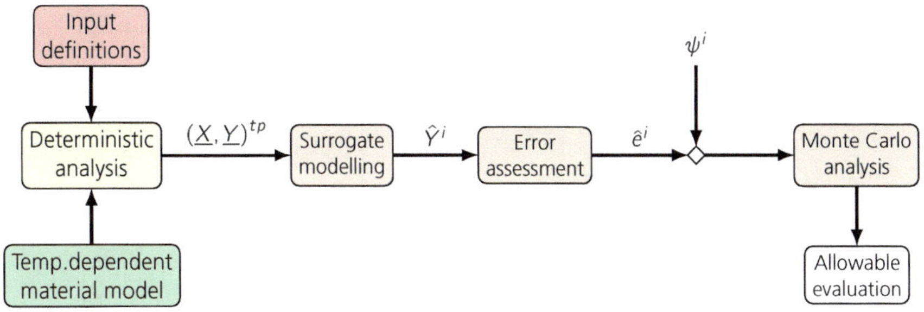

Fig. 5.16 Schematic drawing highlighting the DiA procedure and its individual steps

or other surrogate modelling strategies may be used for improving the prediction quality. This is a crucial point to consider, as the designer must ultimately decide whether the structural behavior and model accuracy are acceptable. Finally, a Monte Carlo analysis is performed to generate a statistical distribution of the design values. These steps highlighted by brown boxes refer to the probabilistic analysis steps and serve as the basis for deriving the final design value allowables. These allowables represent the structural behavior for the defined distributed temperature conditions and their variations.

In addition to the schematic view presented in Figs. 5.16, 5.17 illustrates the analysis procedures and their outcomes across the parameter space using a schematic example. The input parameters that describe the modelling space are shown along the x and y axes as the average temperature (T^{avg}) and the maximum temperature (T^{max}), respectively. They represent the temperature distribution and typically result from thermal analysis of FEA, i.e., $X = \{T^{avg}; T^{max}\}$. The shape of the temperature distribution might be described by more than these two parameters and this parameter reduction used here is further explained later in the scope of structural analysis assessment in Chap. 6. The red point in the figure represents the uniform analysis result, which serves as a reference point and is located at a temperature level $T^{uni} = T^{max} = T^{avg}$. The black dashed horizontal line represents the maximum temperature, which is commonly designated as the temperature limit or maximum service temperature. This temperature limit is also addressed later in the probabilistic analysis as it defines the maximum hotspot temperature. The solid black line in Fig. 5.16 represents all uniform solutions of different temperatures where $T^{avg} = T^{max}$. Since the temperature distributions resulting from an external heat source are being investigated, there is a local hotspot. Therefore, all analyses related to distributed temperature conditions must be above this line since $T^{max} > T^{avg}$. This is a natural condition when assuming that the temperature overshoot occurs only in part of the structure. The black points in Fig. 5.17 represent the results of deterministic analyses under different thermal conditions, i.e. training points. Each training point corresponds to a pair of values for Xth and is related to a specific design value within Y_j. These training points are distributed within the parameter space and are located above the solid black line but not restricted by the temperature

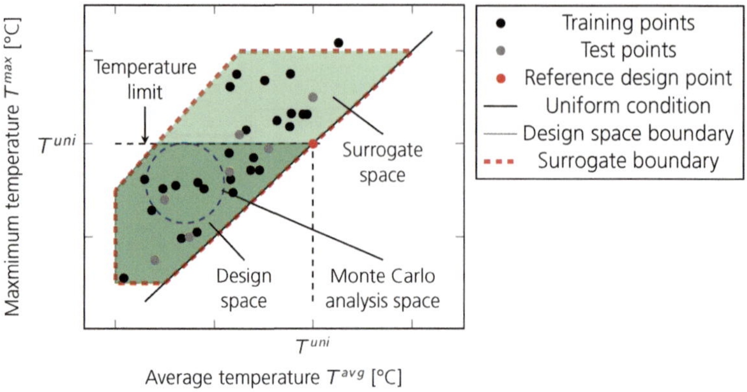

Fig. 5.17 Exemplaric overview to analysis procedure results and the addressed input parameter space highlighting the difference between analysis of uniform temperature (UnA) and distributed temperature (DiA)

limit to generate sufficient model precision in this domain. Based on the training points, surrogate models are derived, with each design value represented by its own surrogate model. The green model space within the red-dashed border represents the surrogate space that is applicable to the problem of distributed temperature. However, to avoid extrapolation errors, the surrogate space considered for probabilistic analysis is reduced to a domain with good training point resolution and, therefore, some training points might fall outside. This domain is determined based on the distribution of the training points while the points outside the parameter space are still used to train the model. After constructing the surrogate models, error assessment is conducted using the LOO approach and the test approach. The test approach uses additional points shown in gray and these points are not part of the training point dataset, which is used for LOO error assessment. This assessment helps to ensure that the surrogate models accurately capture the behavior of the underlying system.

The final step of the analysis procedure is the Monte Carlo analysis, which generates a large number of data points Xth using surrogate models to determine the design values $\hat{Y}_j$. The surrogate space used in this step is limited to the parameter space below the temperature limit, which defines the worst-case thermal condition, and is represented by a darker green visualization. The resulting Monte Carlo data points are distributed within the blue dashed circle, which describes the joint statistical distribution of the input parameters that can actually occur, including the worst-case scenario approaching the temperature limit. The resulting set of design values depends on the statistical distribution introduced for the input parameters within Xth. Evaluating the resulting design value distribution leads to the allowables, which represent the structural behavior for distributed temperature conditions. These allowables are the result of the explained DiA procedure and can be compared to the related UnA result point shown in red.

This approach is applied in the following chapter to selected examples, which are used to discuss and compare the distributed temperature approach to the uniform temperature analysis. By using these examples, the effectiveness and accuracy of the approach can be evaluated and potential advantages and limitations can be identified. The comparison to the uniform temperature analysis allows for a better understanding of the differences and benefits of the distributed temperature approach.

5.6 Summary and Conclusion

This chapter presents the analysis methodology developed in this work and is designated as DiA procedure. It allows for determining the mechanical behavior of structures under spatially distributed thermomechanical load conditions and considering variations of the thermal conditions. The analysis methodology uses the temperature dependent material models developed in Chap. 4 and addresses the main aspects needed to investigate Working Hypotheses 2, 3, and 4. It thereby contributes to the research objective. The DiA procedure includes deterministic FEA, design value assessment, and probabilistic analysis. By following this procedure, the design value allowables are obtained, which represent the structural performance under distributed thermal load conditions. DiA procedure and resulting allowables represent the novelty developed with this work and are applicable to common structural design procedures, as these allowables can replace the deterministic design values obtained using uniform thermal conditions.

The deterministic FEA procedure used is a sequential analysis strategy that consists of thermal analysis, thermal deformation analysis, and mechanical load analysis. The applied thermal load is generated by a parametric function that defines a certain temperature distribution. The material models developed within Chap. ?? are applied to determine the structural behavior at these distributed thermal conditions. The linear relation between the surface values of global strain and temperature is used for calculating the elastic strain components and assessing material failure in dependence on temperature. This approach can also be used to assess experimental data, which is a valuable outcome for future work on validation. However, the FEA procedure considers the stress-free temperature and a thermoelastic approach to take into account residual stress resulting from manufacturing. These stresses can become severe and significantly contribute to the failure condition. Furthermore, these residual stresses and the effective thermomechanical laminate behavior depend on the laminate's setup and the actual temperature. The structural behavior captured by these analyses is assessed with respect to buckling and material failure (IFF and FF). The buckling assessment is performed by evaluating the stiffness evolution in dependence of applied mechanical load. Failure is evaluated using various state-of-the-art criteria, such as proposed by Hashin, Puck, and Cuntze. Overall, the FEA procedure accounts for mentioned dependencies of temperature and temperature distribution and creates the basis for a probabilistic analysis of composite structures under varying thermomechanical load conditions and thus contributes to investigating the reseach obective.

A surrogate modelling procedure converts the discrete design values for buckling and material failure, determined at discrete thermal conditions using the FEA procedure, into a function of temperature distribution parameters. The selected modelling approach is a Kriging method. The resulting surrogate models are evaluated for their accuracy using both leave-one-out (LOO) and test-dataset methodologies and compared against the precision of the FEA method to assess the surrogate accuracy. Once the surrogate function is determined, a Monte-Carlo analysis is conducted efficiently using the surrogate to obtain the final allowables, representing the structural performance at different thermal conditions. These allowables can then be compared to the current design values, which are based on uniform conditions.

References

1. H.D. Baehr, K. Stephan, Wärme- und Stoffübertragung. 3. Au?. Springer-Lehrbuch. Berlin and Heidelberg: Springer (1998). ISBN: 3-540-64458-X. http://swbplus.bsz-bw.de/bsz069142815cov.htm
2. E.A. Thornton, Thermal buckling of plates and shells. Appl. Mech. Rev. **46**(10), 485–506 (1993). https://doi.org/10.1115/1.3120310
3. M. Liebisch et al., Robustness analysis of CFRP structures under thermomechanical loading including manufacturing defects. Athens (2019). https://www.sucohs-project.eu/publications-presentations
4. M. Liebisch, T. Wille, Exploitation of high performance structures at elevated tempera-tures, in *Aerospace Europe Conference 2020* (2020)
5. R. Degenhardt et al., Design and analysis of stiffened composite panels including post- buckling and collapse. Comput. & Struct. **86**(9), 919–929 (2008). ISSN: 00457949. https://doi.org/10.1016/j.compstruc.2007.04.022
6. D. Wilckens, Beul- und Nachbeulverhalten von versteiften CFK-Schalen unter Druck- und Schubbelastung. Dissertation. Braunschweig: TU Braunschweig, 2018
7. C. Mittelstedt, W. Becker, Strukturmechanik ebener Laminate. 1. Au?age. Darmstadt (2016). ISBN: 978-3-935868-99-0
8. F.R. Jones, M. Mulheron, J.E. Bailey, Generation of thermal strains in GRP: Part 2 The ori-gin of thermal strains in polyester cross-ply laminates. J. Mater. Sci. **18**, 1533–1539 (1983). https://doi.org/10.1007/BF01111974 [TitelanhanddieserDOIinCitavi- Projektbernehmen]
9. J. Magnus Svanberg, Predictions of manufacturing induced shape distortions: high performance thermoset composites. Doctoral Thesis. Lulea, Sweden: Lulea University of Technology, 2002. https://www.diva-portal.org/smash/record.jsf?pid=diva2%3A998979&dswid=2428
10. R. Hein, Vorhersage und In-Situ Bewertung fertigungsbedingter Deformationen und Eigenspan-nungen von Kompositen. Dissertation. TU Braunschweig, 2019. https://elib.dlr.de/128886/
11. R. Hein, T. Wille, M. Liebisch, Probabilistic process simulation and in-situ process prediction during composite manufacturing as contribution to industry 4.0, in *NAFEMS World Congress 2019, Quebec City, Canada* (2019). https://elib.dlr.de/130811/
12. M. Liebisch, R. Hein, T. Wille, FE-based prediction of process induced distortions and residual stresses for cfrp frames, in *5th EASN International Workshop on Aerostructures, Manch-ester, UK* (2015)

13. M. Liebisch, R. Hein, T. Wille, Probabilistic process simulation to predict process induced distortions of a composite frame. CEAS Aeronaut. J. **9**(4), 545–556 (2018). ISSN:1869-5582. https://doi.org/10.1007/s13272-018-0302-7
14. H.T. Hahn, N.J. Pagano, Curing stresses in composite laminates. J. Compos. Mater. **9**(1), 91–106 (1975). https://doi.org/10.1177/002199837500900110
15. Y. Weitsman, Residual thermal stresses due to cool-down of epoxy-resin composites. J. Appl. Mech. **46**(3), 563–567 (1979). ISSN: 00218936. https://doi.org/10.1115/1.3424606
16. R. Prussak et al., Smart cure cycles for fiber metal laminates using embedded fiber Bragg grating sensors. Compos. Struct. 213, 252–260 (2019). ISSN: 02638223. https://doi.org/10.1016/j.compstruct.2019.01.079
17. E. Kappel, R. Prussak, J. Wiedemann, On a simultaneous use of fiber-Bragg-gratings and strain-gages to determine the stress-free temperature Tsf during GLARE manufacturing. Compos. Struct. 227, 111279 (2019). ISSN: 02638223. https://doi.org/10.1016/j.compstruct.2019.111279
18. P. Prasatya, G.B. McKenna, S.L. Simon, A viscoelastic model for predicting isotropic residual stresses in thermosetting materials: effects of processing parameters. J. Compos. Mater. **35**(10), 826–848 (2001). ISSN: 0021-9983. https://doi.org/10.1177/a037322
19. C. Brauner, Analysis of process-induced distortions and residual stresses of composite structures. Dissertation. Universität Bremen. ISBN: 978-3-83-25-3528-5
20. S.P. Timoshenko, J.M. Gere, Theory of Elastic Stability. International Student Edition, Second Edition (1961)
21. J. Singer, J. Arbocz, T. Weller, *Buckling Experiments: Experimental Methods in Buckling of Thin-Walled Structures: Shells, Built-Up Structures, Composites and Additional Topics* (John Wiley and Sons, Chichester and New York, 2002). ISBN: 9780471974505. https://doi.org/10.1002/9780470172995
22. L. Liu et al., Thermal buckling of heat-exposed axially restrained composite column. Compos. Part A **37**, 972–980 (2006). https://doi.org/10.1016/j.compositesa.2005.04.006
23. L.-W. Chen, L.-Y. Chen, Thermal buckling of laminated composite plates. J. Therm. Stress. **10**(4), 345–356 (1987). ISSN: 0149-5739. https://doi.org/10.1080/01495738708927017
24. L.-W. Chen and L.-Y. Chen, Thermal buckling of laminated cylindrical plates. Compos. Struct. **8**(3), 189–205 (1987). ISSN: 02638223. https://doi.org/10.1016/0263-8223(87)90069-9
25. C.A. Meyers, M.W. Hyer, Thermal buckling and postbuckling of symmetri-cally laminated composite plates. J. Therm. Stress. **14**(4), 519–540 (1991). ISSN: 0149-5739. https://doi.org/10.1080/01495739108927083
26. L.-W. Chen, L.-Y. Chen, Thermal buckling analysis of composite laminated plates by the finite-element method. J. Therm. Stress. **12**(1), 41–56 (1989). ISSN: 0149-5739. https://doi.org/10.1080/01495738908961953
27. A.K. Noor et al., Thermomechanical buckling and postbuckling of multilayered composite panels. Compos. Struct. **23**(3), 233–251 (1993). https://doi.org/10.1016/0263-8223(93)90225-F
28. H.-S. Shen, Z.-Q. Lin, Thermal post-buckling analysis of imperfect laminated plates. Comput. & Struct. **57**(3), 533–540 (1995). ISSN: 00457949. https://doi.org/10.1016/0045-7949(94)00628-G
29. H.-S. Shen, G. Sun, F.W. Williams, Thermomechanical postbuckling analysis of imperfect laminated plates on two-parameter elastic foundation. Compos. Struct. **34**(3), 325–338 (1996). ISSN: 02638223. https://doi.org/10.1016/0263-8223(95)00153-0
30. H.-S. Shen, F.W. Williams, Thermal postbuckling analysis of imperfect laminated plates on softening non-linear elastic foundations. Compos. Sci. Technol. **57**(7), 719–728 (1997). ISSN: 02663538. https://doi.org/10.1016/S0266-3538(97)00027-4
31. H.-S. Shen, Thermomechanical post-buckling analysis of imperfect laminated plates using a higher-order shear-deformation theory. Comput. & Struct. **66**(4), 395–409 (1998). ISSN: 00457949. https://doi.org/10.1016/S0045-7949(97)00089-8
32. H.-S. Shen, Thermal postbuckling behavior of imperfect shear deformable laminated plates with temperature-dependent properties. Comput. Methods Appl. Mech. Eng. **190**(40), 5377–5390 (2001). https://doi.org/10.1016/S0045-7825(01)00172-4

33. H.-S. Shen, Q.S. Li, Thermomechanical postbuckling of shear deformable laminated cylindrical shells with local geometric imperfections. Int. J. Solids Struct. **39**(17), 4525–4542 (2002). https://doi.org/10.1016/S0020-7683(02)00351-7
34. M.K. Singha, L.S. Ramachandra, J.N. Bandyopadhyay, Thermal postbuckling analy-sis of laminated composite plates. Compos. Struct. **54**(4), 453–458 (2001). ISSN:02638223. https://doi.org/10.1016/S0263-8223(01)00117-9
35. L.-W. Chen, L.-Y. Chen, Thermal buckling behavior of laminated composite plates with temperature-dependent properties. Compos. Struct. **13**(4), 275–287 (1989). ISSN: 02638223. https://doi.org/10.1016/0263-8223(89)90012-3
36. L.-W. Chen, L.-Y. Chen, Thermal postbuckling analysis of laminated composite plates by the finite element method. Compos. Struct. **12**(4), 257–270 (1989). ISSN: 02638223. https://doi.org/10.1016/0263-8223(89)90075-5
37. A.K. Noor, W.S. Burton, Three-dimensional solutions for the thermal buckling and sensitivity derivatives of temperature-sensitive multilayered angle-ply plates. J. Appl. Mech. **59**(4), 848 (1992). ISSN: 00218936. https://doi.org/10.1115/1.2894052
38. A.R. Vosoughi et al., Thermal postbuckling of laminated composite skew plates with temperature dependent properties. Thin Walled Struct. **49**(7), 913–922 (2011). ISSN: 02638231. https://doi.org/10.1016/j.tws.2011.02.017
39. J.-S. Chang, A further study on thermal buckling of simply supported antisymmetric angle-ply laminates in a unifrom-temperature field. Compos. Sci. Technol. **43**(4), 309–315 (1992). https://doi.org/10.1016/0266-3538(92)90054-7
40. A.K. Noor, W.S. Burton, Computational models for high-temperature multilayered composite plates and shells. Appl. Mech. Rev. **45**(10), 419 (1992). ISSN: 00036900. https://doi.org/10.1115/1.3119742
41. R. Zimmermann, H. Klein, A. Kling, Buckling and postbuckling of stringer stiffened fibre composite curved panels—tests and computations. Compos. Struct. **73**(2), 150–161 (2006). ISSN: 02638223. https://doi.org/10.1016/j.compstruct.2005.11.050
42. B. Kriegesmann, Probabilistic design of thin-walled fiber composite structures. Dissertation. Gottfried Wilhelm Leibniz Universität Hannover: Hannover, 2012. https://www.repo.uni-hannover.de/handle/123456789/7944
43. J. Singer, J. Arbocz, T. Weller, *Buckling Experiments: Experimental Methods in Buckling of Thin-Walled Structures: Basic Concepts, Columns, Beams and Plates* (John Wiley and Sons, Chichester and New York, 2002). ISBN: 9780471956617. https://doi.org/10.1002/9780470172988
44. L.K. Chang, M.F. Card, Thermal buckling analysis for stiffened orthotropic cylindri-cal shells: NASA-TN-D-6332 (1971). https://ntrs.nasa.gov/citations/19710014258
45. R. Rolfes, S. Tacke, R. Zimmermann, Development and experimental verification of FE-model for stringer-stiffened fibre composite panels under combined thermal and mechanical loading conditions, in *European Conference on Spacecraft Structures, Materials and Mechanical Testing*, Braunschweig (1998), pp. 109–115. https://www.dlr.de/fa/Portaldata/17/Resources/dokumente/institut/1999/s1999_2.pdf
46. N. Breivik, M. Nyer, J. Starnes, Thermal and mechanical buckling and postbuckling responses of selected curved composite panels, in *38th Structures, Structural Dynamics, and Materials Conference* (American Institute of Aeronautics and Astronautics, Reston, Virginia, 1997). https://doi.org/10.2514/6.1997-1247
47. M. Liebisch, Entwicklung von Metamodellen für ausgewählte Auslegungskriterien von or-thotrop versteiften Schalen aus kohlenstofffaserverstärktem Kunststoff. Diplomarbeit. 2014:TU Braunschweig. https://elib.dlr.de/95591/
48. S. Freund, Metamodelle mehrerer Genauigkeitsstufen zur Stabilitätsanalyse versteifter Schalenaus Faserverbundwerkstoffen. Dissertation. TU Braunschweig, 2018. https://elib.dlr.de/137590/
49. P.D. Soden, A.S. Kaddour, M.J. Hinton, Recommendations for designers and re-searchers resulting from the world-wide failure exercise, in *Failure Criteria in Fibre-Reinforced- Polymer Composites* (Elsevier, 2004), pp. 1223–1251. ISBN: 9780080444758. https://doi.org/10.1016/B978-008044475-8/50039-1

50. M. Hinton, A. Kaddour, P. Soden, A comparison of the predictive capabilities of current failure theories for composite laminates, judged against experimental evidence. Compos. Sci. Technol. **62**(12–13), 1725–1797 (2002). ISSN: 02663538. https://doi.org/10.1016/S0266-3538(02)00125-2
51. M. Hinton, A. Kaddour, P. Soden, A further assessment of the predictive capabilities of current failure theories for composite laminates: comparison with experimental evidence. Compos. Sci. Technol. **64**(3–4), 549–588 (2004). ISSN: 02663538. https://doi.org/10.1016/S0266-3538(03)00227-6
52. M.J. Hinton, A.S. Kaddour, P.D. Soden, Evaluation of failure prediction in composite laminates: background to 'part C' of the exercise. Compos. Sci. Technol. **64**(3–4), 321–327 (2004). ISSN: 02663538. https://doi.org/10.1016/S0266-3538(03)00216-1
53. M.J. Hinton et al., *Failure Criteria in Fiber Reinforced Polymer Composites: Can any of the Predictive Theories be Trusted?* (Boston, 2011)
54. A.S. Kaddour et al., The background to the third world-wide failure exercise. J. Compos. Mater. **47**(20–21), 2417–2426 (2013). ISSN: 0021-9983. https://doi.org/10.1177/0021998313499475
55. M.J. Hinton, A.S. Kaddour, The background to part b of the second world-wide failure exercise: evaluation of theories for predicting failure in polymer composite laminates under three-dimensional states of stress. J. Compos. Mater. **47**(6–7), 643–652 (2013). ISSN: 0021-9983. https://doi.org/10.1177/0021998312473346
56. A.S. Kaddour, M.J. Hinton, Maturity of 3D failure criteria for fibre-reinforced composites: comparison between theories and experiments: part B of WWFE-II. J. Compos. Mater. **47**(6–7), 925–966 (2013). ISSN: 0021-9983. https://doi.org/10.1177/0021998313478710
57. A.-S. Kaddour, M.J. Hinton, 1.23 failure criteria for composites, in *Comprehensive Composite Materials II* (Elsevier, 2018), pp. 573–600. ISBN: 9780081005347. https://doi.org/10.1016/B978-0-12-803581-8.10354-6
58. R. Cuntze et al., Neue Bruchkriterien und Festigkeitsnachweise für unidirektionalen Faserkunst-stoffverbund unter mehrachsiger Beanspruchung - Modellbildung und Experimente: Fortschrit-tberichte VDI. 1997
59. J. Echaabi, F. Trochu, R. Gauvin, Review of failure criteria of fibrous composite materials. Polym. Compos. **17**(6), 786–798 (1996). https://doi.org/10.1002/pc.10671
60. A. Puck, Failure analysis of frp laminates by means of physically based phe-nomenological models. Compos. Sci. Technol. **58**(7), 1045–1067 (1998). ISSN: 02663538. https://doi.org/10.1016/S0266-3538(96)00140-6
61. R.G. Cuntze, The predictive capability of failure mode concept-based strength criteria for multi-directional laminates-part B. Compos. Sci. Technol. **64**(3–4), 487–516 (2004). ISSN: 02663538. https://doi.org/10.1016/S0266-3538(03)00225-2
62. R.G. Cuntze, A. Freund, The predictive capability of failure mode concept-based strength criteria for multidirectional laminates. Compos. Sci. Technol. **64**(3–4), 343–377 (2004). ISSN: 02663538. https://doi.org/10.1016/S0266-3538(03)00218-5
63. P. Zinoviev, The strength of multilayered composites under a plane-stress state. Compos. Sci. Technol. **58**(7), 1209–1223 (1998). ISSN: 02663538. https://doi.org/10.1016/S0266-3538(97)00191-7
64. S.W. Tsai, E.M. Wu, A general theory of strength for anisotropic materials. J. Compos. Mater. **5**(1), 58–80 (1971). ISSN: 0021-9983. https://doi.org/10.1177/002199837100500106
65. S.W. Tsai, J.D.D. Melo, An invariant-based theory of composites. Compos. Sci. Technol. **100**, 237–243 (2014). ISSN: 02663538. https://doi.org/10.1016/j.compscitech.2014.06.017
66. A. Puck, *Festigkeitsanalyse von Faser-Matrix-Laminaten: Modelle für die Praxis* (Hanser, München and Wien, 1996). ISBN: 3-446-18194-6
67. Z. Hashin, Failure criteria for unidirectional fiber composites. J. Appl. Mech. **47**, 329–334 (1980). ISSN: 00218936. https://doi.org/10.1115/1.3153664
68. S.P. Timoshenko, I.M. Lessells, *Festigkeitslehre* (Verlag von Julius Springer, Berlin, 1928)
69. A. Puck, H. Schürmann, Failure analysis of FRP laminates by means of physically based phenomenological models. Compos. Sci. Technol. **62**(12–13), 1633–1662 (2002). ISSN: 02663538. https://doi.org/10.1016/S0266-3538(01)00208-1

70. A. Puck, J. Kopp, M. Knops, Guidelines for the determination of the parameters in Puck's action plane strength criterion. Compos. Sci. Technol. **62**(3), 371–378 (2002). ISSN: 02663538. https://doi.org/10.1016/S0266-3538(01)00202-0
71. R.G. Cuntze, Fracture-type strength criteria formulated by invariants which consider the material symmetries of the isotropic/anisotropic material used. Noordwijk. https://articles.adsabs.harvard.edufull/1996ESASP.386..399C/0000401.000.html
72. R.G. Cuntze, Efficient 3D and 2D failure conditions for UD laminae and their application within the veri?cation of the laminate design. Compos. Sci. Technol. **66**(7–8), 1081–1096 (2006). ISSN: 02663538. https://doi.org/10.1016/j.compscitech.2004.12.046
73. R. Cuntze, The predictive capability of failure mode concept-based strength conditions for laminates composed of unidirectional laminae under static triaxial stress states. J. Compos. Mater. **46**(19–20), 2563–2594 (2012). ISSN: 0021-9983. https://doi.org/10.1177/0021998312449894
74. W. Michaeli, M. Mannigel, F. Preller, On the effect of shear stresses on the fibre failure behaviour in CFRP. Compos. Sci. Technol. **69**(9), 1354–1357 (2009). ISSN: 02663538. https://doi.org/10.1016/j.compscitech.2008.09.024
75. E.C. Edge, Does transverse and shear loading affect the compression strength of uni- directional CFC? A reply to Dr Hart-Smith. Composites **25**(2), 159–161 (1994). ISSN:00104361. https://doi.org/10.1016/0010-4361(94)90010-8
76. A.G. Gibson et al., High temperature and fire behaviour of continuous glass fibre/polypropylene laminates. Compos. Part Appl. Sci. Manuf. **41**(9), 1219–1231 (2010). ISSN: 1359835X. https://doi.org/10.1016/j.compositesa.2010.05.004
77. S.T. Pinho et al., Material and structural response of polymer-matrix fibre-reinforced composites. J. Compos. Mater. **46**(19–20), 2313–2341 (2012). ISSN: 0021-9983. https://doi.org/10.1177/0021998312454478
78. S. Feih et al., Modelling the tension and compression strengths of polymer laminates in fire. Compos. Sci. Technol. **67**(3–4), 551–564 (2007). ISSN: 02663538. https://doi.org/10.1016/j.compscitech.2006.07.038
79. S. Feih et al., Modelling the compression strength of polymer laminates in fire. Compos. Part Appl. Sci. Manuf. **38**(11), 2354–2365 (2007). ISSN: 1359835X. https://doi.org/10.1016/j.compositesa.2007.04.013
80. A.G. Gibson et al., Modeling composite high temperature behavior and fire response under load. J. Compos. Mater. **46**(16), 2005–2022 (2011). ISSN: 0021-9983. https://doi.org/10.1177/0021998311429383
81. R. Gutkin et al., Micro-mechanical modelling of shear-driven fibre compressive failure and of fibre kinking for failure envelope generation in CFRP laminates. Compos. Sci. Technol. **70**(8), 1214–1222 (2010). ISSN: 02663538. https://doi.org/10.1016/j.compscitech.2010.03.009
82. P. Prabhakar, A.M. Waas, Interaction between kinking and splitting in the compressive failure of unidirectional fiber reinforced laminated composites. Compos. Struct. **98**, 85–92 (2013). ISSN: 02638223. https://doi.org/10.1016/j.compstruct.2012.11.005
83. M. Bishara, R. Rolfes, O. Allix, Revealing complex aspects of compressive failure of polymer composites—part I: fiber kinking at microscale. Compos. Struct. **169**, 105–115 (2017). ISSN: 02638223. https://doi.org/10.1016/j.compstruct.2016.10.092
84. G. Balokas et al., Probabilistic analysis of composite stiffened panels including random AFP-produced defects towards enhanced design allowables. Compos. Struct. 116923 (2023). ISSN: 02638223. https://doi.org/10.1016/j.compstruct.2023.116923
85. A.I.J. Forrester, A. Sóbester, A.J. Keane, *Engineering Design via Surrogate Modelling: A Practical Guide* (Wiley, 2008). ISBN: 978-0-470-06068-1. http://www.loc.gov/catdir/enhancements/fy0829/2008017093-d.html

Chapter 6
Structural Analysis and Assessment

This chapter utilizes the novel Distributed temperature Analysis procedure (DiA) procedure, which applies distributed temperature conditions and is described in Chap. 5. In that respect, the temperature dependent A-value material allowable models, developed in Chap. 4, are applied. The structural behavior and the design values, including Local Buckling (LB), Inter-Fiber Failure (IFF), and Fiber Failure (FF), are analyzed using DiA and they are compared to the common deterministic analysis procedure that applies uniform thermal conditions (abbreviated as Uniform temperature Analysis procedure (UnA)). These comparisons lead to an assessment of the analysis procedure and resulting structural potentials, described at the end of this chapter. The investigations are conducted on an unstiffened, flat shell structure, which is representative for a skin field, conf. Figure 5.4. This structure is expected to exhibit all relevant and critical effects that may arise for such structural elements under thermomechanical load conditions with distributed thermal load and that were found in previous studies on stiffened structures [1, 2]. By conducting the investigations on this simplified example instead of a stiffened structure, better understanding of the resulting effects can be achieved leading to a suitable assessment of the overall DiA methodology.

6.1 Test Matrix

This chapter utilizes the DiA methodology to compare it with the current deterministic UnA procedure for the skin field test structure shown in Fig. 5.4. To evaluate the deviations, advantages, and limitations of the two analysis procedures, specific test cases with related variations are defined. Furthermore, three general questions

M. Liebisch, *Design Load Allowables for Composite Plates Exposed to Thermomechanical Loads*, Mechanics and Adaptronics,
https://doi.org/10.1007/978-3-031-71352-1_6

Table 6.1 Overview to test case variations. Each line represents a test case and is evaluated regarding both statistical scenarios

Designation	Variation of laminate stacking		Variation of thermal load case	
	t_{lam} (mm)	Laminate stacking (°)	$T_{max} = T_{uni}$ (°C)	Temperature range (°)
CP-A	3	$[0\|90]_S$	170	140–180
CP-B	3	$[0\|90]_S$	220	195–230
CP-C	3	$[0\|90]_S$	120	90–130
MAP-A	3	$[0\| \pm 45]_S$	170	140–180
MAP-B	3	$[0\| \pm 45]_S$	220	200–230
MAP-C	3	$[0\| \pm 45]_S$	120	90–130

are established to guide the framework of this analysis. These questions will serve as a reference point throughout the chapter's discussion.

1. How does a non-uniform temperature distribution affect the **structural behavior** in comparison to a uniform temperature distribution? What sensitivities exist, and are there any critical deviations?
2. What are the limitations or disavantages of the **modelling and analysis procedure** proposed by the DiA methodology? Can they be minimized or avoided by future advancements?
3. What is the extent of the load-carrying **potential** gained? Can these be **transferred** to more complex structures?

To address these questions, three general types of variations are defined: variation of laminate stacking, variation of thermal load cases, and variation of statistical scenarios. These cases are intended to showcase common scenarios and typical sensitivities, while also capturing the nature of worst-case effects that can occur in combined thermomechanical loading conditions. Table 6.1 presents and overview to laminate stacking and thermal conditions used in deterministic analysis. The scope of statistical scenarios is visualized in Fig. 6.1. These variations are explained in detail below:

Variation of the Laminate Stacking

The behavior of a structure is influenced by its laminate setup. When exposed to changes in temperature, structures with different laminate stackings exhibit varying behavior. To investigate related sensitivities, two laminate setups have been selected: the Cross-Ply setup (CP) and the Modified Angle-Ply Modified Angle-Ply laminate (MAP) setup. These setups are chosen as they are representative of severe cases and that are also used earlier discussing the stiffness behavior over temperature (Fig. 4.8) and the residual stress development (Fig. 5.8) for varying laminate setups.

1. The Cross-Ply laminate (CP) setup is a cross-ply setup ($[0|90]_S$). It has equal effective properties in the x and y directions and shows significant residual stress in transverse ply directions.
2. The MAP setup consists of a $[0|\pm45]_S$ stacking. It exhibits different effective laminate properties in the x and y directions, including varying stiffness and thermal expansion behavior. Additionally, the maximum residual stress developed is much higher in the fiber direction and less critical in the transverse direction, compared to the CP setup.

Variation of the Thermal Load Case

The thermal load case domain is critical for the overall behavior of a structure, as it scales the magnitude of thermal preload, residual stress and material properties. To study the resulting sensitivities on the structural behavior, three different thermal load cases are investigated for both laminates. Each thermal load case is designated by a capital letter. In each case, a reference temperature is defined representing the maximum service temperature T^{max} of a design scenario, which corresponds to T^{uni} and the temperature applied in UnA. The related temperature range highlighted in Table 6.1 refers to the temperature domain that is applied for the set of deterministic analysis of DiA to generate the training points as visualized by the black points in Fig. 5.16. The top bound of this temperature range exceeds T^{uni} and is selected to provide a good surrogate accuracy in the temperature design space up to T^{max}.

1. Case A: The reference temperature of 170 °C is close to the stress-free state and represents a typical service temperature related to the Polyetersulfone (PES) T_g
2. Case B: The reference temperature of 220 °C, equal to the PES T_g, represents a high-temperature condition and consists of severe property degradation. It is beyond current temperature limits.
3. Case C: The reference temperature of 120 °C causes less property degradation and a different residual stress magnitude.

The distribution of temperature and the nature of applied variations is explained in Sect. 5.1.1 and exemplaric temperature distributions are shown in the Appendix, see Fig. C.8.

Variation of Statistical Scenario Applied for the Thermal Load Distribution

The variability of thermal loads is mostly unknown or relatively uncertain. Depending on the location of a structure and the type of thermal impingement, different temperature conditions and different variability in these conditions can occur. Therefore, two different scenarios of statistical distributions are investigated. They are based on equal surrogate models and applied in the scope of the Monte Carlo Analysis. These two scenarios are illustrated in Fig. 6.1 representing two different domains of the parameter space, i.e. the location of the most probable condition, and they refer to different types of variation expressed horizontal and vertical width and resulting distribution. These statistical scenarios are artificially created to showcase the principle of the DiA methodology and to study how thermal load uncertainty manifests within the structural allowables. However, both scenarios are created to consider T^{max} as top bound of temperature.

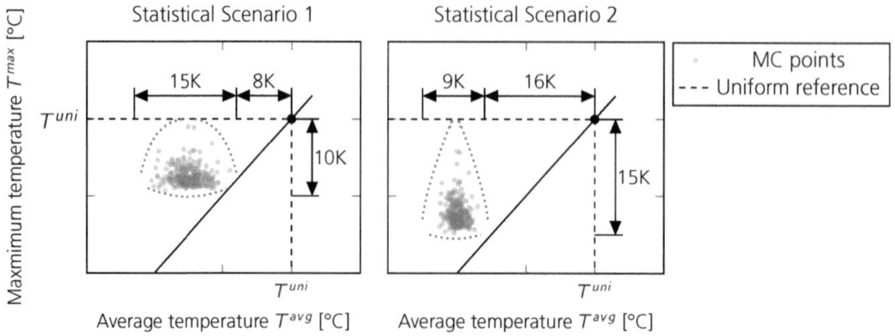

Fig. 6.1 Statistical scenarios applied in Monte Carlo Analysis. This Figure is reduced to 100 points while Monte Carlo Analysis applies 10000 points. The absolute temperature ranges are similarly applied to every thermal case using T^{uni} for reference. The dotted lines indicate the asymmetric shape of the distribution function regarding T^{max}

1. Scenario 1: A relatively steady thermal load case with relatively low range of maximum temperature variation and a relatively wide area of average temperature.
2. Scenario 2: A peak load case, with relatively small variation of average temperature but a relatively high local temperature overshoot.

Comparison and Potential Evaluation

The comparison between the two analysis approaches is conducted for each design criterion and the related design values Y^j refer to the axial force F_{ax}^j applied to the test structure. The final evaluation is done by comparing the deterministic analysis result at uniform reference condition (UnA) to three characteristic design values obtained in distributed thermal conditions by the DiA procedure. The most conservative design value is the minimum design load resulting:

$$F_{ax}^{min} = min\left(\underline{F_{ax}}^j\right) \tag{6.1}$$

Additional design values are determined using the typical allowable margins providing F_{ax}^A and F_{ax}^B, respectively. To compare between the analysis approaches, the mechanical potential is expressed by the ratio of *A-value* and uniform reference as percentage:

$$\Delta^P\left(Y^i\right) = \left(\frac{\hat{Y}^{i,A}}{Y^{i,uni}} - 1\right) \tag{6.2}$$

Remark on Analysis Settings

The thermal start condition of Finite Element Analysis (FEA) is sensitive to the mechanical load that is generated by thermal expansion and the clamped conditions at the boundaries. It is equally defined in both, UnA and DiA and the resulting load is similarly existent in both procedures. This load magnitude is designated as the thermal preload, F_{ax}^{u0}, as it is caused within the thermal deformation analysis step prior to mechanical load analysis. Therefore, the thermal start condition is the most simple parameter that can be tuned and is well comprehensible, since it scales the thermal preload. In common use, this temperature would be defined by Room Temperature (RT) as the condition, where the structure is assembled and mechanical Boundary Condition (BC) are enforced. However, in the present analysis, the thermal start condition is selected equal to the stress-free temperature of 160 °C, which is a realistic magnitude as explained in Sect. 5.1.4. This adaption is made, as it considers the relation between residual stress and thermal stress development. It generates specific effects that result to the variations discussed and the related sensitivities are referred to again at the related point within the discussions. As indicated in Sect. 5.1, a thermal analysis FEA is substituted by a direction specification of the temperature distribution. Exemplaric temperature distributions applied are shown in the Appendix Fig. C.8. The analysis procedure and boundary conditions applied are explained in Sect. 5.1 and within the Appendix Sect. C.1.

6.2 Outline to Result Evaluation

The following sections discuss and explain the results of the test cases described within the previous section. Specifically, Sect. 6.3 focuses on the analysis results of the CP structure, while Sect. 6.4 presents analysis results of the MAP structure. Both explain the sensitivities resulting for thermal load case variation and the statistical scenarios at the respective laminates. This section serves as an introduction to these sections, providing readers with a basis for understanding and interpreting the results.

The research objective of this thesis includes the development of a structural analysis methodology that considers distributed thermal conditions and their load variation and that has the potential to replace the common analysis procedure which applies uniform conditions. In Chap. 5, the developed DiA methodology is explained, and the results presented in the following sections are generated using this methodology. To compare the results the DiA procedure with the uniform analysis methodology, designated as UnA, it should be noted that the UnA always refers to a single temperature, while the DiA procedure contains variations of temperature distributions. These variations are summarized through the thermal parameters T^{avg} and T^{max}.

For both structures, the assessment and comparison of results between UnA and DiA is divided into:

- Deterministic Analysis and
- Probabilistic Analysis

The Deterministic Analysis Section focuses on the structural mechanical observations and sensitivities. The section begins with an overview to the structural analyses performed and the results are explained on plots that visualize the relation between design values and T^{max}. Firstly, the load-shortening curves and design values obtained from the UnA analyses are presented. The design values exhibit a specific progression over temperature, which is assessed in comparison to a linear trend. Arrows highlight the reference values of the thermal load case. Secondly, the load-shortening curves and design values obtained from the DiA analyses are presented in a similar figure. In both figures Blue, green, and red markers indicate the design value results for LB, IFF, and FF, respectively. The second chart, showing the results of the DiA methodology, contains the comparison between UnA results and DiA results. Therein, following details are considered:

- The comparison is conducted for the three thermal cases A, B and C
- The load-shortening curves are separated horizontally to distinguish the thermal cases. This is only done for differentiate the results for better visualization
- The UnA results are shown by the thermal reference cases and are drawn as black squares
- The design value trends of the UnA results are highlighted by black-dashed lines to compare the scattering data obtained by the DiA analyses as result to the thermal load variations considered

The Probabilistic Analysis Section focuses on the accuracy assessment of the surrogate models and the evaluation of structural potentials. Both quantities are evaluated by different metrics and presented using a single table. The surrogate models are constructed using the deterministic DiA results. The accuracy of the surrogate models is evaluated by comparing the estimated error magnitudes against the analysis uncertainty quantity ψ, conf. Sect. 5.4. The final design value results and potentials are compared at the bottom of the table and are visualized in a separate figure. This includes a chart for each design value and thermal case. In these charts, the UnA results are represented by red vertical lines. Both statistical cases are evaluated showing the outcome of the DiA procedure. They are represented by blue and green color, respectively and obtain a certain range of design values shown by a histogram on which the final design value allowables are derived (vertical lines). The difference between the UnA and DiA lines represents the potentials which can also be found in the table.

6.3 Cross-Ply Laminate

6.3.1 Deterministic Analysis

Uniform Temperature Analysis (UnA) Results

Figure 6.2 illustrates the results of the UnA analyses for the CP structure. The left chart displays all load-shortening curves determined for temperatures in the range of 85–225 °C. The design values determined in these analyses, i.e. LB, IFF, and FF, are shown by blue, green, and red markers, respectively. The right chart of Fig. 6.2 presents these design values over temperature applying equal y-axis scaling as for the load-shortening curves. The black markers in the right chart highlight the thermal preload that results from thermal expansion and refers to the y-axis on the right side. The colored dashed lines indicate the trends related to each design value in similar

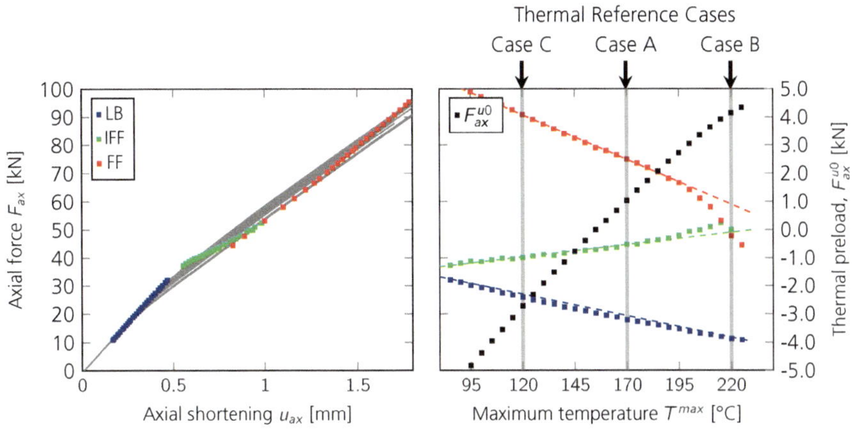

Fig. 6.2 Deterministic FEA results for the CP laminate at various uniform temperatures

colors. Buckling exhibits an almost linearly decreasing trend over the full temperature range. The magnitude of the buckling load reduction is similar to the change of thermal preload created. This correlation is expected and this part of load refers to a contribution of thermal buckling margin while thermal and mechanical strains are superposed here. IFF displays an increasing trend, which becomes progressive at around 200 °C. FF demonstrates a linear decreasing trend similar to buckling, but with a more pronounced decrease from 200 °C onwards. These observations of IFF, and FF trends are explained in more detail below.

Distributed Temperature Analysis (DiA) **Results**

Figure 6.3 displays the results of the DiA analyses, with similar colors used for the design values as in the previous figure. In the top part of the figure, three charts are presenting the load-shortening curves obtained by DiA in gray and each chart refers to a single thermal load case. A single load-shortening curve is plotted in black in each of these charts highlighting the corresponding reference case obtained by UnA along with the related design values shown as black markers. As in the previous figure, the chart located in the bottom presents the design values as a function of temperature, with similar coloring used for DiA. In the scope of UnA, the maximum temperature, shown on the x-axis, is equal to the average temperature while for the DiA procedure, the average temperature is lower due to the nature of temperature distribution, as explained in Sect. 5.5. The black dashed lines represent the trends from UnA (conf. Fig. 6.2) and are useful for highlighting the differences between UnA and DiA results. The arrows and gray vertical lines indicate the reference temperature of the individual thermal cases, highlighting the selected UnA reference values drawn for comparison with DiA. Qualitatively similar trends are observed between UnA and DiA and a detailed comparison and analysis between UnA and DiA results is described below.

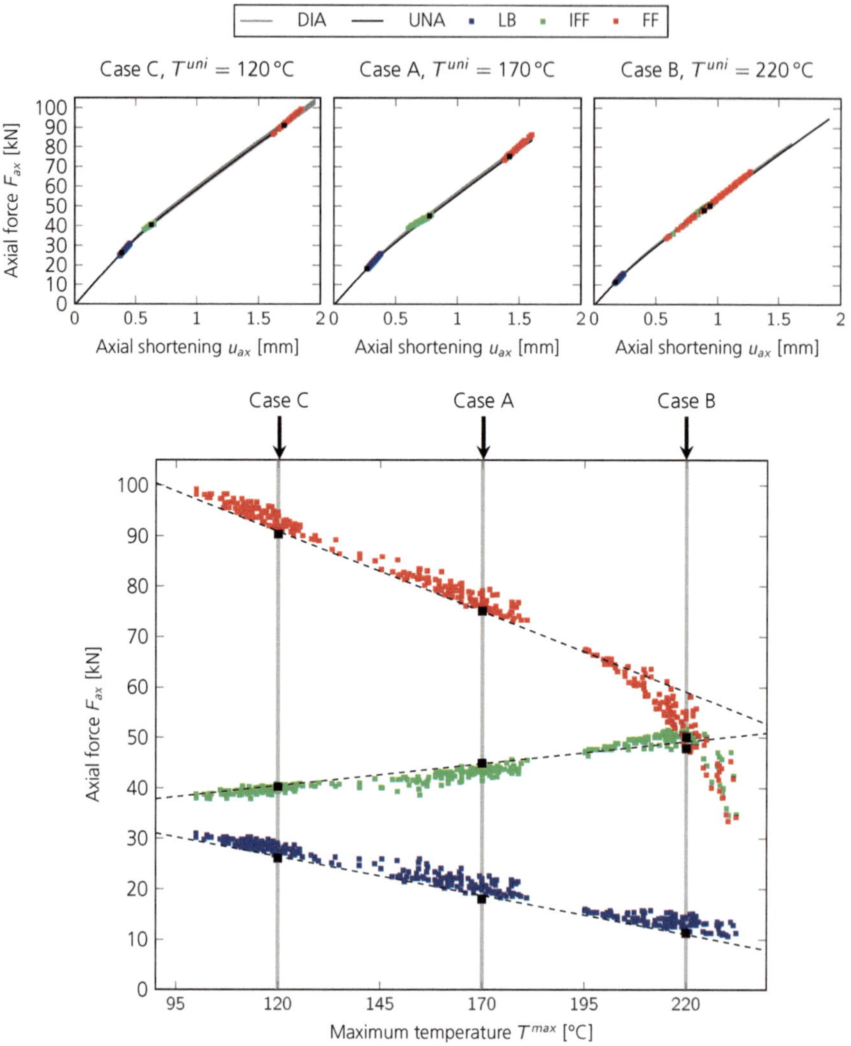

Fig. 6.3 Deterministic FEA results indicating the structural behavior of the CP laminate for the three thermal cases: 120 °C (**C**), 170 °C (**A**) and 220 °C (**B**). Uniform analysis results are shown for the related reference temperatures with design values as black squares, respectively. The figure presents a reduced number of load-shortening curves but all design values determined

Comparison: Buckling Behavior and Deformation Behavior
A similar deformation is observed in all thermal cases for both UnA and DiA, as exemplarily shown in Fig. 5.4. In the initial state, i.e., the prebuckling state, the structural stiffness is approximately 70 $\frac{\text{kN}}{\text{mm}}$ and there are no significant deflections. Normal deflections resulting in DiA as thermal deformation are distributed similarly to the respective temperature shapes. Their magnitude correlates with temperature

and is below 5% of the laminate thickness across all cases. In conclusion, these deformations are insensitive to buckling here and are therefore not further discussed. The postbuckling deformation describes a buckling pattern with three half-waves. During buckling, the structural stiffness follows a relatively smooth S-shape towards a postbuckling stiffness of about 47 $\frac{\text{kN}}{\text{mm}}$. The structural stiffness is similarly sensitive to temperature in both, prebuckling and postbuckling. This is expected and in line with the study discussed in Fig. 4.4. For the sake of clarity, the progression of the structural stiffness across all thermal cases is provided in the appendix (see Fig. C.1).

The buckling loads exhibit a decreasing trend with temperature elevation. This is due to the reduction of effective stiffness properties and the thermal preload, adds compressive loads for increasing temperature. The LB trend that corresponds to the UnA results, i.e. the black dashed line in Fig. 6.3, represents the lower bound of the DiA buckling loads. The scatter of DiA values above this trend is mainly due to the combination of two factors that act in superposition. First, the temperature distribution leads to a lower average temperature in DiA and results in a higher effective structural stiffness compared to UnA. Second, the lower magnitude of effective thermal expansion in DiA leads to a lower thermal preload. Although the thermal deformation could potentially act as resistance against buckling towards the natural buckling shape, causing the higher force in LB, this has not been verified. Also, these deformations do not act as sensitive imperfection in the present case.

Comparison: Material Failure Behavior

Across all cases and for both criteria, IFF and FF, failure is predicted to occur within the postbuckling regime. As a result, the deformation state at the initiation of failure is similar across all cases. This observation holds true for both UnA and DiA analyses. To illustrate the differences between the two analysis approaches, a comparison of strain and failure indices are shown in Figs. C.2 and C.3, respectively, for an exemplary case. Further explanation to this figure follows after discussing the behavior of FF and IFF.

Compressive FF is the dominant failure mode observed in all cases, with a linearly decreasing trend over temperature up to 200 °C, followed by a degressive trend at higher temperatures. This trend is similar to the correlation between temperature and compression strength in the fiber direction. The contribution of load redistribution from the 90° plies into the 0°-plies is likely negligible in both approaches. However, in DiA, higher strength properties are observed in distance to the hotspot due to the lower temperature compared to the uniform case. This is a sensitive effect, as compression strength is particularly dependent on temperature. The higher FF loads determined in DiA compared to UnA are mainly attributed to this observation. The sensitivity of residual stress contribution is small, conf. Figure 5.8. Similarly, the thermal preload contribution to the difference in FF loads between UnA and DiA is small, as the variation of thermal preload in the investigated temperature ranges is small compared to the margin of FF, see Fig. 6.2.

The failure criterion of IFF exhibits an increasing trend of failure loads with increasing temperature. The dominant failure mode observed is the transverse tensile failure mode. As illustrated in Fig. 5.8, the thermal stress in the CP laminate transverse to the fibers is tensile at temperatures below the stress-free condition ($T^{\sigma 0} = 160\,°\mathrm{C}$), and becomes compressive above this temperature. In parallel, the thermal stress (i.e. the thermal preload) exhibits a similar trend. The sensitivity of failure to these thermal and residual stresses is higher compared to the sensitivity that exists from the transverse tensile strength in this particular case and causes the increasing trend over temperature. This is true until about 220 °C, which represents the T_g of PES and is a critical temperature for the material behavior. Above this temperature, the trend changes to a degressive behavior where the material strength behavior becomes more sensitive. However this is out of scope for the present analysis. The described observations are consistent for both analysis procedures.

The trend of DiA results proceeds mostly below the UnA values. This is for a similar reason as for FF, which is the structural behavior in distance to the hotspot. Following the explanation above, sections distant to the hotspot, colder areas, contain higher tensile strength compared to the center locations in UnA. However, As residual and thermal stress are predominant, this leads to lower IFF forces caused in areas distant from the center.

Conclusion on Structural Behavior
The differences between UnA and DiA conditions lead to positive load carrying potentials for LB and FF. For local buckling, these potentials are a result of high effective properties and lower thermal preload. In case of FF, the dominant sensitivity for these potentials is obtained by the location of failure initiation and the predominant material strength. A similar effect between both analysis procedures is observed for IFF. However, here, thermal and residual stress effectively act different compared to FF causing lower values for IFF in UnA compared to DiA. This observation is critical, since it reveals, that the simplifications made by UnA are not conservative regarding IFF and overestimate this behavior. A more detailed analysis approach as provided by the DiA can attribute these deviations.

The sensitivity that results from the selected starting conditions (see end of Sect. 6.1) causes an initial offset in axial force. Adapting this condition changes the axial preload, indicated in Fig. 6.2. Noteworthy is, that such preload exists similarly in both, x and y direction. Starting conditions of lower temperature, e.g. RT, result in additional compressive force and stress, i.e. negative values of F_{ax}^{u0}. Hence, such changes reduce the design value by the related offset. Depending on the magnitude, buckling can even occur as thermal buckling and within the thermal deformation analysis step. As a result to the nature of the CP laminate investigated here, the transverse load resulting to such thermal preload is not critical. However, this can become the case for different laminate setups or structures.

6.3.2 *Probabilistic Analysis*

Table 6.2 presents an overview to the accuracy of the surrogate models and the design value results determined for the CP laminate variations. The columns of the table

Table 6.2 Overview to surrogate model settings, surrogate accuracy and design value results determined for the CP laminate. The surrogate model errors are determined on surrogates using all available training points (quantity n). Positive and negative signs of MaxAE indicate if the model error occurs in positive or negative direction, relative to the surrogate, conf. Eq. 5.28

	Thermal case		Case C, n = 120			Case A, n = 122			Case B, n = 122		
	Criterion j		LB	IFF	FF	LB	IFF	FF	LB	IFF	FF
Surrogate evaluation	$X_1^{th} = T^{avg}$	(°C)	{92.4; 127.0}			{121.2; 168.5}			{183.5; 225.2}		
	$X_2^{th} = T^{max}$	(°C)	{100.0; 135.0}			{140.0; 181.0}			{195.2; 232.5}		
	$min\left(\psi^j\right)$	(kN)	0.30	0.82	0.75	0.24	0.64	0.62	0.24	0.63	0.62
	$max\left(\psi^j\right)$	(kN)	0.30	0.95	0.75	0.29	0.79	0.74	0.29	0.76	0.75
	$\bar{\psi}^j$	(kN)	0.30	0.88	0.75	0.26	0.68	0.64	0.26	0.66	0.66
	Error assessment using test dataset										
	$MaxAE^j$	–	(+)	(+)	(−)	(−)	(+)	(−)	(−)	(−)	(−)
	$MaxAE^j$	(kN)	0.19	1.51	0.99	0.28	1.88	1.23	0.71	2.92	3.20
	$RMSE^j$	(kN)	0.11	0.57	0.41	0.13	0.65	0.49	0.11	0.58	0.71
	Error assessment using LOO approach										
	$MaxAE^j$	–	(+)	(−)	(−)	(−)	(−)	(+)	(−)	(−)	(+)
	$MaxAE^j$	(kN)	0.71	1.65	6.09	0.63	1.72	3.90	0.71	2.90	4.30
	$RMSE^j$	(kN)	0.16	0.50	0.79	0.19	0.73	0.73	0.16	0.76	1.18
Design value determination	Uniform analysis										
	$F_{ax}^{j,T^{uni}}$	(kN)	26.0	40.2	90.9	18.0	44.9	75.1	11.3	50.1	47.8
	Distributed analysis—Statistical scenario 1										
	$F_{ax}^{j,min}$	(kN)	27.2	37.6	92.4	19.4	42.4	76.1	12.3	49.3	51.0
	$F_{ax}^{j,A}$	(kN)	27.8	39.0	93.4	19.7	42.5	77.0	12.6	49.4	54.0
	$F_{ax}^{j,B}$	(kN)	28.1	39.3	93.9	20.0	42.8	77.4	12.9	49.7	55.7
	$\Delta^P\left(F_{ax}^{j,A}\right)$	(%)	+6.9	−3.0	+2.8	+9.4	−5.6	+2.5	+11.5	−1.4	+13.0
	Distributed analysis—Statistical scenario 2										
	$F_{ax}^{j,min}$	(kN)	28.7	38.8	94.9	20.3	42.0	77.3	13.4	48.8	54.7
	$F_{ax}^{j,A}$	(kN)	29.0	39.1	95.5	20.7	42.1	78.5	13.6	48.9	58.4
	$F_{ax}^{j,B}$	(kN)	29.1	39.2	95.8	20.9	42.3	79.2	13.7	48.9	59.8
	$\Delta^P\left(F_{ax}^{j,A}\right)$	(%)	+11.5	−2.7	+5.1	+15.0	−6.5	+4.5	+20.3	−2.4	+22.2

represent the individual thermal cases, and the corresponding design values. The top part of the table presents the surrogate model evaluation. Each surrogate model is created and evaluated for the related training point space defined by X^{th}. The accuracy is rated using the analysis uncertainty, ψ as indicated in Eq. 5.27. This value can vary from analysis to analysis and depends on the FEA incrementation used. It is therefore expressed through the range obtained by all analyses performed and the resulting average value. The errors are determined as explained in Sect. 5.4, using the test dataset evaluation procedure and the LOO evaluation procedure, respectively. The bottom part of the table compares the design values obtained from the UnA and the results analyzed by the DiA procedure. The DiA results distinguish between the two statistical scenarios investigated. For each scenario, three different margins of design load allowables are presented and compared. The resulting load carrying potential Δ^P between UnA and DiA is expressed in relation to the A-value allowables mentioned.

Surrogate Model Evaluation

Overall, all surrogate models display good accuracy. In most cases, Root Mean Square Error (RMSE) values are either below or close to the chosen analysis uncertainty value of ψ. One exception is observed by FF in Case B, where both the test data and the LOO approaches show RMSE values above the analysis uncertainty. In most cases, the MaxAE values are above the analysis uncertainty, indicating high local deviations and in some cases, the surrogate models exhibit high sensitivity to individual training points, as indicated by high MaxAE values for the LOO approach, which is especially the case for FF. However, it is unclear whether these errors are captured in the center of the surrogate space or at the outer area where extrapolation errors can occur. Severe errors are highlighted by positive signs representing overestimation of the model compared to the test data. The evaluations of the test dataset methodology presents critical values for IFF in case A and case C. The error found in case C for LB is uncritical, as its magnitude falls below the bound of the accuracy criterion.

The surrogate models are constructed using all the available training points and the training points and surrogate model representations are visualized in Appendix Sect. C.3. However, some regions of the parameter space have a higher density of training points compared to others. Such uneven distribution of training points arises from the definition of the thermal input parameters, which are finally represented by only two parameters T^{avg} and T^{max}. However, the use of all parameters needed to describe the temperature distribution would lead to a much higher dimensionality and less density of training points. As a result to uneven training point distribution, certain regions of the parameter space may have fewer training points, particularly the outer regions. It is expected that the surrogate models may have higher errors in these areas due to insufficient extrapolation accuracy of the model approach. Such high local errors may negatively impact the global error metric. However, the RMSE values are within the proximity of ψ and therefore, the models are sufficiently precise. This implies that the surrogate accuracy could be even better at the inner space of the model domain, which is of particular interest for the statistical analysis and proposes to satisfy the accuracy criterion. Thus, it would be advisable to assess the surrogate quality only within the area of interest and not in the outer regions where training points are rare. Although training points in outer regions are important for building a reliable surrogate model, they may be excluded from the evaluation of model accuracy. This could potentially help for a more reliable surrogate evaluation.

Probabilistic Analysis and Potential Evaluation

Figure 6.4 illustrates the results of the probabilistic analysis for the three different thermal cases analyzed, respectively. Each chart shows a histogram that represents the statistical distribution of the design values, with blue and green indicating statistical scenario 1 and scenario 2, respectively. The design values, i.e., LB, IFF, and FF, are distinguished through separated plots that are horizontally aligned from left to

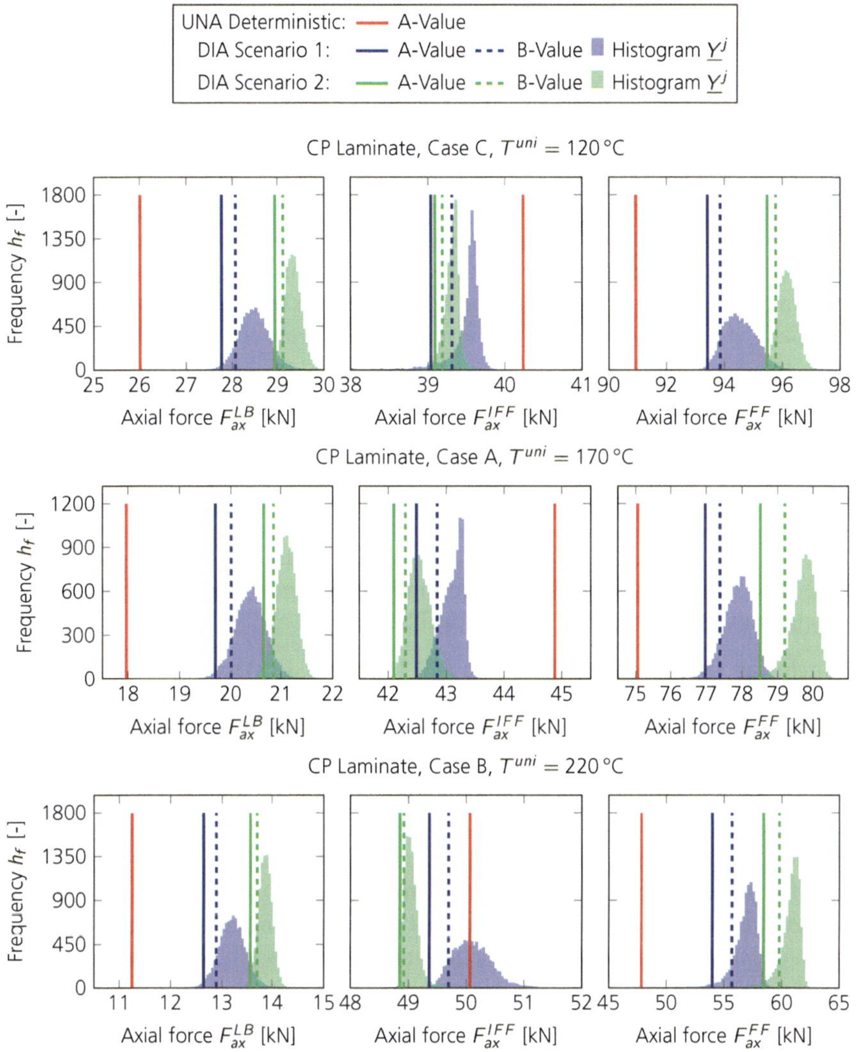

Fig. 6.4 Design values for the CP Laminate. Comparison between UnA results (red) and DiA results where blue and green indicate statistical scenarios 1 and 2, respectively, and using solid and dashed lines to indicate A-values and B-values

right. The vertical lines highlight the actual design values, which are also listed in Table 6.2. The dashed and solid green and blue lines indicate the *B-value* and *A-value* margins of the related DiA result, respectively. The red line represents the design value obtained by the UnA. The difference between the blue and red solid lines, as well as between the green and red solid lines, represents the load potential obtained by the DiA procedure for the two statistical scenarios, which is indicated by Δ^P in Table 6.2.

The results confirm the sensitivities captured from the deterministic results, showing that positive load-carrying potential exists in all cases for LB and for FF. Negative load-carrying potential is observed for IFF. Moreover, the statistical analysis allows to quantify the potentials by comparing the UnA result in red to the DiA results which correspond to specific cases of temperature distribution. For buckling, the difference in axial force between UnA and DiA is roughly constant over temperature while the percentual potential increases due to the decreasing magnitude of LB. The lower integral thermal strain that exists in distributed temperature conditions is the dominating effect for this, which is similar over all thermal cases as the absolute temperature variations are similar. For FF, such axial force difference increases especially at high temperature while also being relatively constant in Case A and C. Here, the governing effect is the material strength degradation in combination with the failure location, which is less severe in distributed conditions and distant from the hotspot. In all thermal cases, the load-carrying potential captured by the DiA procedure is higher for statistical scenario 2 compared to statistical scenario 1 which is consistent with the previous observations, as the average temperature of statistical scenario 2 is lower. A similar but negative trend occurs for IFF. At low temperature, the difference between UnA and DiA is higher and reduces towards higher temperature. The statistical scenario also affects this dependency, with Scenario 2 showing a higher mismatch compared to Statistical Scenario 1.

Conclusion on Surrogate Modelling and Probabilistic Analysis

The CP laminate structure example represents a desirable design case for applying the DiA procedure. This includes satisfying the requirement for surrogate modelling which requests consistent structural behavior over the input parameter space. In fact, for the CP test case, the deformation behavior of both approaches, UnA and DiA, show consistent deformation behavior within the full thermal parameter space.

The resulting surrogate models are generated with sufficient accuracy and are used for probabilistic analysis to derive structural allowables which represent the behavior in relation to the specified thermal load variation, here represented through the statistical cases. A certain demand for improving the surrogate model error quantification methodology is derived. However, the reasons to this are not of systematic nature and highlight the importance of thorough error discussion. The comparison of results verifies that load-carrying potentials exist in structures sized by the UnA procedure. Further the DiA procedure confirms to be effective in identifying these potentials and in generating useful quantities of structural allowables which may serve in a more global design process. The comparison of two different statistical scenarios highlights the sensitivity resulting to thermal load variation and the ability of the DiA procedure to be applied in varying thermal conditions. This variation also emphasizes an additional advantage of the DiA procedure, since it can become a useful tool if utilized for sensitivity analysis.

6.4 Modified Angle-Ply Laminate

The CP laminate example, discussed in the previous section, demonstrates a desirable design case which satisfies the surrogate modelling requirement of consistent deformation. However, the present section describes the behavior of the MAP laminate example which is not a desirable case but rather an example designed to illustrate both the limitations and benefits of the DiA procedure. Therefore, the explanation of the MAP example follows a similar structure as used for the CP example but highlights the challenges and limitations that can arise for specific structures and thermal load conditions.

6.4.1 *Deterministic Analysis*

Uniform Temperature Analysis (UnA) Results

Figure 6.5 displays the results of the UnA analyses for the MAP structure, which are conducted in a similar manner as for the previous CP-laminate example. In the left chart, all load-shortening curves from 75 to 220 °C are presented, with colored markers indicating the design values. The right chart shows the design values drawn over temperature using the same y-axis, with trends indicated by colored dashed lines. In addition, the thermal preload is indicated by black markers using the y-axis on the right hand side.

It can be observed that the load-shortening curves exhibit diverging trends, with varying slopes and buckling load variations. As indicated in the right chart, the trends of LB and FF follow similar degressive patterns as for the CP laminate. However,

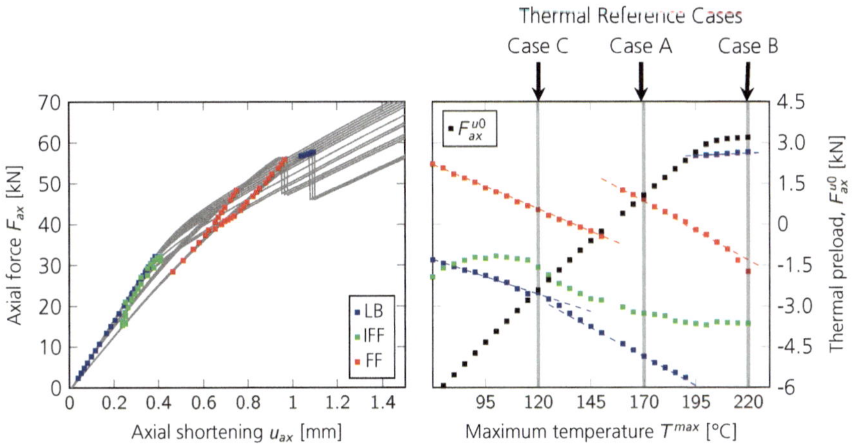

Fig. 6.5 Deterministic FEA results indicating the structural behavior of the MAP laminate for varying uniform temperatures

a discontinuity is apparent for FF around 160 °C which represents the magnitude of the stress-free temperature used. These observations represent variations in the structural behavior and the reasons for these trends are discussed in detail below. However, it should be noted that the thermal preload relation over temperature does not show such discontinuities.

Distributed Temperature Analysis (DiA) Results

Figure 6.6 displays the results obtained for the MAP structure using the DiA approach, which have been processed in a similar way as explained before. The load-shortening curves are shown in three separate charts at the top of the figure, each one representing a thermal load case, and the colored markers present the design values evaluated. In these charts, the black load-shortening curve represents the UnA results at reference temperature and the black markers refer to the related design values, serving as comparison to the DiA results. The bottom part of the Fig. 6.6 presents the design value relations over temperature, again comparing the deterministic DiA results to the trends observed for the UnA analyses which are marked by dashed lines. The gray vertical lines highlight the reference temperatures of the thermal load cases.

While the curves for thermal cases B and C show a consistent behavior within their respective cases, the curves for case A exhibit a clear divergence. Some of them follow a trend similar to case B, while others resemble more closely the curves from case C. This behavior is caused by the buckling response of the structure which changes in the range of case A. These changes lead to different deformation patterns. Case C follows a similar shape as observed for the CP case. In Case A, some curves follow a similar trend, while others clearly diverge and show a snap-through (negative stiffness) at higher loads. In Case B, all curves proceed similarly to the changed curve that has the snap-through. Consequently, these pecularities lead to specific trends in the design values as presented by Fig. 6.6 and will be explained in more detail below.

The relative behavior between UnA and DiA is similar compared to the CP laminate. The trend lines obtained by UnA and the design values show similar progression. The UnA trend lines represent a lower bound to the DiA values of LB and FF which was also observed for the CP laminate. The behavior of IFF over temperature is not following a linear trend which makes it hard to compare. However, also these values are following a similar relation compared to the UnA results.

This particular example represents an undesired design case. The results show that the load-carrying potential of the structure is highly sensitive to temperature changes. This observance is caused by the different laminate properties in the x and y directions. The combination of thermal expansion and stiffness behavior depending on the preferential direction lead to a structural behavior that is highly sensitive to temperature. The temperature, at which this sensitivity occurs, is in this case setup through the initial conditions of the simulation. The causes and effects of this sensitivity to temperature are discussed in more detail in the following paragraphs. Understanding and addressing such sensitivities are crucial in ensuring that the design is robust and can withstand various environmental conditions.

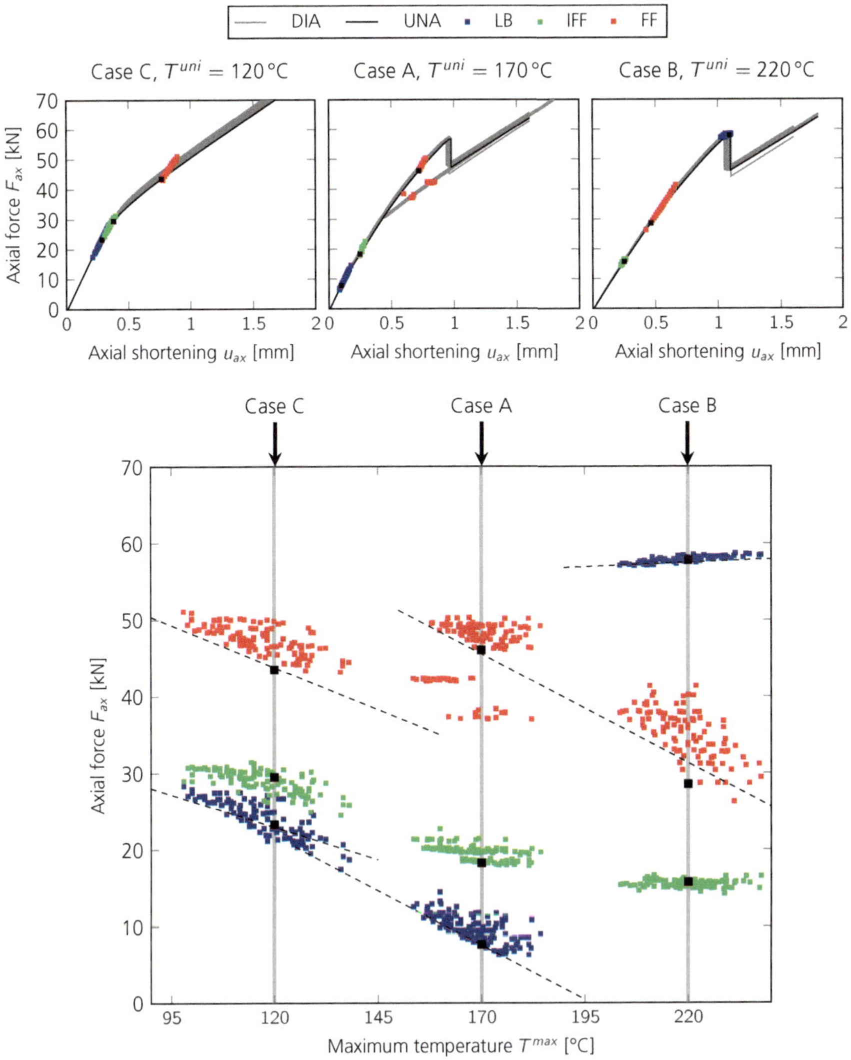

Fig. 6.6 Deterministic FEA results indicating the structural behavior of the MAP laminate for the three thermal cases: 120 °C (**C**), 170 °C (**A**) and 220 °C (**B**). Uniform analysis results are represented at reference temperatures with design values as black squares, respectively. The load-shortening curves distinguish the three example cases by a horizontal offset and are represented by 10 curves per case

Comparison: Deformation Behavior and Buckling

The MAP laminate exhibits varying buckling and deformation behavior in response to temperature changes due to its effective laminate properties which depend on the load direction. This is illustrated by the load-shortening curves discussed above and shown in Fig. 6.6. To understand the underlying structural mechanical effects responsible for these deviations, the deformation states that arise for the uniform analysis reference of Case A are discussed first. Figure 6.7 illustrates the characteristic deformation states observed, which also represent all principal deformation variations observed for other thermal load cases of the MAP structure. In addition to that, the development of stiffness over axial displacement is a useful relation to present the occurring effects. The progression of structural stiffness over axial load is exemplarily shown in Fig. 6.8 for selected temperature distributions and the related temperature distributions are attached within the Appendix, conf. Fig. C.8. The highlighted points or sections of the stiffness-shortening curves correspond to the deformation states shown in Fig. 6.7 and the colors indicate the related thermal load case. The blue dashed line in the Fig. 6.8 represents the uniform analysis result of Case A at 170 °C. After discussing this particular analysis result, the deviations observed in other thermal cases are explained.

Under uniform thermal conditions at 170 °C (Case C, blue dashed line in Fig. 6.8), no deflection occurs without additional loading, as indicated in (i). This state represents the prebuckling state and corresponds to a structural stiffness of approximately 87 $\frac{\text{kN}}{\text{mm}}$. As axial compression is applied, the structure deforms in a single direction, as shown in (ii). This longitudinal deflection is similar to the natural thermal deformation that occurs under distributed thermal conditions and causes a drop in stiffness to around 70 $\frac{\text{kN}}{\text{mm}}$. The transition from (i) to (ii) is triggered by the compressive stress state in the y-direction that is a result to thermal expansion. With further axial load increase, the magnitude of deflection increases and the structural stiffness decreases (iii). This superposed state of membrane and bending stress, present from (ii) to (iii), acts effectively as resistance against the buckling towards the common postbuckling shape consisting of three half-waves (iv). However, at increasing axial load, i.e. loading in x-direction, such a shape is reached through buckling resulting in a load-drop and negative structural stiffness. The final deformation state (iv) is a stable postbuckling deformation with a structural stiffness of around 29 $\frac{\text{kN}}{\text{mm}}$.

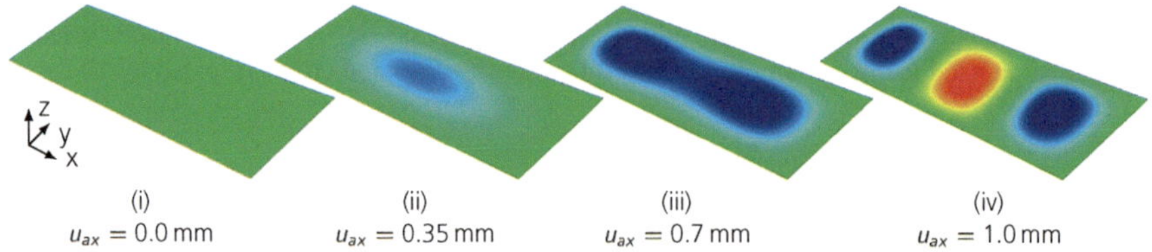

Fig. 6.7 Out-of-plane deformation (z-direction) resulting to axial loading for the MAP laminate at 170 °C uniform temperature. The related stiffness behavior is shown in Fig. 6.8 by the blue dashed line

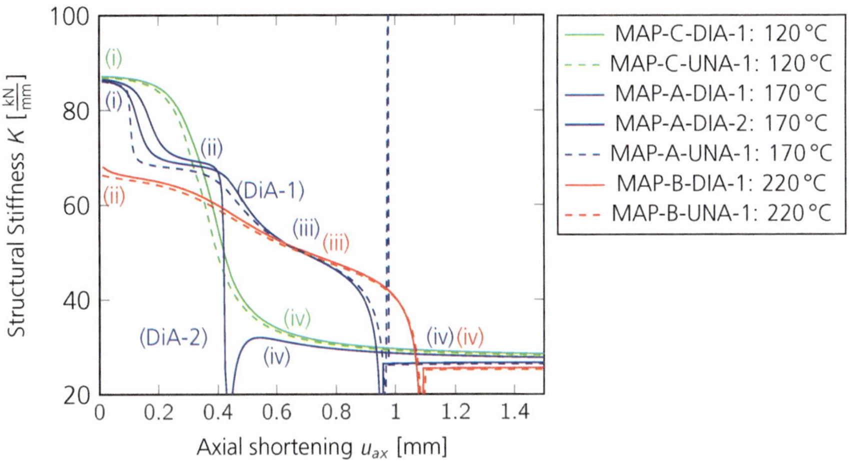

Fig. 6.8 Comparison of the MAP laminate stiffness progression in dependence to axial load. Thermal load cases are indicated by colors. Distributed temperature conditions are represented by solid lines, uniform thermal conditions by dashed lines. The markings (i) to (iv) refer to the actual deformation state in reference to Fig. 6.7. The related temperature distributions are shown within the appendix Fig. C.8

In Case B, shown as the red curves in Fig. 6.8, higher temperatures result in greater thermal expansion, causing the transition from (i) to (ii) to occur during the thermal deformation step and without additional axial loading. This change in deformation represents thermal buckling and is, again, caused by thermal expansion in y-direction. Consequently, the curves start at deformation state (ii) and a lower stiffness level. When axial compression loads are applied, the deformation behavior covers the transitions to (iii) and (iv), as previously explained but in a slightly different stiffness progression over load. Uniform thermal condition and distributed temperatures lead to comparable behavior.

In Case C, highlighted in green, temperatures are evidently below the stress-free temperature, causing thermal expansion to act as tensile strain. The deformation behavior is similar to that observed for the CP laminate. A relatively smooth transition is observed between the prebuckling state (i) and the postbuckling state (iv) without having intermediate (thermal) deformation states as those shown by Fig. 6.7 (ii) and (iii). Once more, uniform and distributed temperatures result in comparable deformation and stiffness reduction.

In addition to Figs. 6.7, 6.8 and C.8 in the Appendix provides a comprehensive overview of the entirety of the structural stiffness curves determined for the MAP laminate. By comparing the cases and the explanation provided above, it is possible to derive valuable sensitivities for understanding the underlying phenomenology that causes the discrepancy obtained in Fig. 6.6.

Temperatures below the stress-free temperature lead to a more stable deformation behavior since the effective stress state resulting from thermal expansion is tensile with less out-of-plane deformation and thus lower sensitivity of geometrical imperfection. On the other hand, thermal expansion in the lateral direction at temperatures above the stress-free temperature causes buckling that is different from the typical one ((i) to (iv)). This can be caused either fully by thermal expansion, as in Case B, and is referred to as thermal buckling, or act as an imperfection to the common axial load case. The resulting buckling state and similar thermal deformation states can also lead to higher buckling loads than the typical buckling loads observed.

In addition to these general conclusions, it is important to analyze the variations that occur within the individual thermal cases by comparing the differences between uniform and distributed temperature conditions, which are represented by the dashed and solid lines of equal color in Fig. 6.8. As mentioned earlier, cases B and C exhibit a comparable curve progression, with the uniform analysis curves mostly below the distributed ones. This trend is attributed to the lower effective laminate property, caused by the temperature differences and also observed for all other thermal conditions in these cases. On the other hand, case A shows distinct deviations, as mentioned earlier. The thermal buckling transition from (i) to (ii) appears smoother for the distributed temperature condition. However, the most significant deviation occurs around $U_{ax} = 0.4$ mm axial shortening, where buckling from (ii) to (iv) takes place, conf. curve DiA-2. This results in buckling loads that are roughly 50% lower than the typical magnitude in this case. Such preemptive buckling behavior is triggered by a specific thermal condition that causes an effective imperfection, which is not found in uniform conditions and is also not observed for every temperature distribution in this thermal load case, as for instance expressed by the curve DiA-1. This phenomenon is critical and causes the variations of the load-shortening curve.

Comparison: Material Failure Behavior

The deformation behavior of the structure significantly influences material failure, as presented in Figs. 6.5 and 6.6 for both IFF and FF. The trend of failure loads for IFF is relatively nonlinear over temperature, which is contrary to the expected linear trend observed in the CP laminate example. This is due to the sensitivity of failure loads to the deformation state. In the case of the CP laminate, the IFF of case C is primarily associated with the onset of buckling, as indicated by overlapping blue (representing onset of buckling) and green squares which explains the observed nonlinear trends. At higher temperatures, particularly for case B, IFF is found to correspond to a distinct deformation state with a low dependency on temperature.

The failure loads of FF demonstrate a decreasing trend with increasing temperature and are also found to be sensitive to the overall deformation state. For case C, FF occurs in the postbuckling regime, while for case B, it is associated with the thermally deformed prebuckling state. The analysis results for case A show a jump in F_{ax}^{FF}, while the distributed analysis results correlate to the variability of the deformation state. Despite this, the trend of FF predicted in uniform analyses is found to be at the lower margin of the distributed analyses.

Conclusion on Structural Behavior

The MAP laminate example highlights the high sensitivity that results from the correlation of laminate stacking setup and temperature to the buckling and deformation behavior of structures in thermomechanical load conditions. Thermal buckling occurs due to relatively high thermal expansion and low stiffness in transverse direction (y), which is observed for both the UnA and the DiA case. By adapting the starting conditions, extreme cases are generated, which help to understand the sensitivities in structural behavior and limitations of the DiA procedure. Case C describes a tensile preload caused by cool-down, case B formulates a load case with thermal buckling, and a snap-through buckling resulting at higher mechanical loads compared to case C. In case A, the thermal preload is compressive but does not lead to thermal buckling. However, specific temperature distributions can lead to preemptive buckling and directly cause a transition to the final postbuckling state, while other thermal states cause much higher buckling forces. Such buckling is caused by the deformation behavior that corresponds to these specific temperature distributions and represents severe imperfections triggering buckling. These imperfections cannot be achieved using the UnA approach. This is a significant advantage of the DiA procedure since it provides a margin of thermal load cases that lead to critical behavior. Although this is not a desired phenomenon, the DiA procedure thereby can support engineers as these studies provide a certain sensitivity study regarding such critical response. Hence it may contribute to identify robust design cases.

The shown extreme cases are generated by adapting the start condition of the temperature. However, similar behavior can also be anticipated in real conditions. The crucial aspect for this is given by the interaction of the effective behavior of structure and the predominant boundary conditions, which finally cause the stress and deformation states.

6.4.2 Probabilistic Analysis

The inconsistent deformation behavior observed in Case A violates the requirement of surrogate modelling for consistent deformation. The stability behavior in this case is a critical example that causes discontinuities in the response function, which is especially for FF design values. These discontinuities make it difficult to reproduce with surrogate modelling, particularly with low number of training points. Therefore, the requirement of continuous deformation behavior is introduced (conf. Sect. 5.4) within the scope of the DiA procedure to avoid error-prone surrogate modelling. As this requirement is not satisfied, surrogate modelling is not conducted for thermal case A, here and the MAP design would not be followed in a design process. It is important to note that such surrogate modelling limitations (low number of training points for complex functions) are not unique to the DiA procedure but rather a general challenge in the field of structural analysis. Nonetheless, the DiA approach could potentially model such problems when increasing the number of deterministic analysis or adapting suitable surrogate modelling techniques.

To investigate the performance of the DiA procedure for the MAP laminate, surrogate modelling and probabilistic assessment are performed for thermal cases B and

C. Compared to the previously described CP laminate, case C exhibits a typical deformation behavior but with different sensitivities due to the changed laminate setup. On the other hand, case B is clearly influenced by the (thermal) deformation behavior. The initiation of IFF and FF occurs in the thermally deformed domain, which has lower structural stiffness and describes a continuous increase in bending deflection. In contrast, LB is obtained here through snap-through bifurcation generating the typical buckling pattern and causing less buckling load variation. Nevertheless, both examples provide a consistent deformation behavior and therefore satisfy the requirements for surrogate modelling. The surrogate models for the thermal cases B and C are constructed using all available training points. An overview of the surrogate models and the corresponding training points is provided in Appendix Sect. C.5.

The probabilistic analysis evaluation of the MAP laminate is conducted following a similar approach as for the CP laminate. The overview to the surrogate accuracy and design value results is presented in Table 6.3. It is again divided into two parts: the upper part presents the surrogate accuracy, while the lower part presents the design value results.

Surrogate Model Evaluation

Overall, the accuracy of the surrogate models is similar to that of the CP laminate example. For the identification of IFF and FF, all RMSE values fall within the range of analysis uncertainty. The magnitude of the MaxAE is mostly within the margin of $2 \cdot \psi^i$, and there are no significant outliers. These observations hold for both error assessment approaches. Regarding buckling, the errors are higher than the analysis uncertainty. However, the absolute values are reasonably small, while the ψ baseline is relatively low in both thermal cases. Hence, the criterion is not met. As previously indicated, an adapted error evaluation that focuses on the parameter space of interest would help to increase the confidence of the surrogate models. Errors in the outer regions could be excluded by such adaptations and propose lower error margins that fulfill the criterion.

Probabilistic Analysis and Potential Evaluation

Figure 6.9 presents the results of the probabilistic analysis for thermal cases B and C using the same coloring and denotations as for the CP laminate.

The buckling behavior of the thermal cases needs to be examined separately to assess their potential. In case C, buckling is caused by the typical smooth occurrence as seen for the CP laminate. The buckling allowable values are noticeably affected by the statistical scenario. For Scenario 1, the allowable buckling loads are almost coincident to UnA, while Scenario 2 indicates potentials of about 9.4%. The histograms of the two scenarios describe different ranges and overlap to some extent. This indicates the dependence of the buckling load on the (tensile) thermal preload, which correlates with the average temperature.

In Case B, buckling occurs in a snap-through bifurcation, indicated by negative stiffness. The design loads of UnA and DiA are close, but DiA yields lower values and negative potentials below 1%. The lower values reflect the sensitivity of thermal deformation to buckling, which is expected in regard to the buckling phenomenon being is likely sensitive to such imperfections. Although this sensitivity is small, it

Table 6.3 Overview to surrogate model settings, surrogate accuracy and design value results determined for the MAP laminate. The surrogate model errors are determined on surrogates using all available training points (quantity n). Positive and negative signs of MaxAE indicate if the model error occurs in positive or negative direction, relative to the surrogate, conf. Eq. 5.28

	Thermal case		Case C, n = 117			Case B, n = 94		
	Criterion j		LB	IFF	FF	LB	IFF	FF
Surrogate evaluation	$X_1^{th} = T^{avg}$	(°C)	{87.7; 127.8}			{188.0; 223.1}		
	$X_2^{th} = T^{max}$	(°C)	{98.0; 138.1}			{203.5; 237.6}		
	$min\left(\psi^j\right)$	(kN)	0.22	0.83	0.43	0.01	0.85	0.70
	$max\left(\psi^j\right)$	(kN)	0.38	1.13	0.49	0.25	1.06	0.88
	$\bar{\psi}^j$	(kN)	0.26	0.97	0.45	0.05	0.93	0.78
	Error assessment using test dataset							
	$MaxAE^j$	–	(+)	(−)	(+)	(+)	(−)	(+)
	$MaxAE^j$	(kN)	1.17	1.62	0.44	0.48	1.16	0.92
	$RMSE^j$	(kN)	0.43	0.81	0.20	0.20	0.51	0.43
	Error assessment using LOO approach							
	$MaxAE^j$	–	(+)	(−)	(+)	(−)	(−)	(−)
	$MaxAE^j$	(kN)	1.99	2.86	0.85	1.17	2.92	1.57
	$RMSE^j$	(kN)	0.56	1.09	0.31	0.35	0.77	0.58
Design Value Determination	Uniform Analysis							
	$F_{ax}^{j,uni}$	(kN)	23.3	29.5	43.5	57.7	15.7	28.5
	Distributed analysis—Statistical scenario 1							
	$F_{ax}^{j,min}$	(kN)	22.7	28.4	45.3	57.3	15.4	32.6
	$F_{ax}^{j,A}$	(kN)	23.4	28.8	46.0	57.4	15.4	34.2
	$F_{ax}^{j,B}$	(kN)	23.8	29.1	46.6	57.5	15.4	35.6
	$\Delta^P\left(F_{ax}^{j,A}\right)$	(%)	+0.4	−2.4	+5.7	−0.5	−1.9	+20.0
	Distributed analysis—Statistical scenario 2							
	$F_{ax}^{j,min}$	(kN)	24.7	29.4	47.3	57.1	15.3	36.5
	$F_{ax}^{j,A}$	(kN)	25.5	29.8	47.7	57.3	15.3	36.9
	$F_{ax}^{j,B}$	(kN)	25.7	29.9	48.0	57.4	15.4	37.2
	$\Delta^P\left(F_{ax}^{j,A}\right)$	(%)	+9.4	+1.0	+9.7	−0.7	−2.5	+29.5

can be concluded that the DiA procedure provides more accurate results as it allows to considers such sensitivities in contrast to the UnA methodology.

The assessment of material failure design loads reveals similar observations to those found for the CP laminate. For IFF, the differences between UnA and DiA are small and often fall below the surrogate error margin. Nevertheless, the trend is consistent with the results for the CP laminate, as the DiA procedure tends to yield slightly lower values compared to UnA, reflecting the sensitivities arising from the more detailed resolution of the temperature field. On the other hand, clear potentials

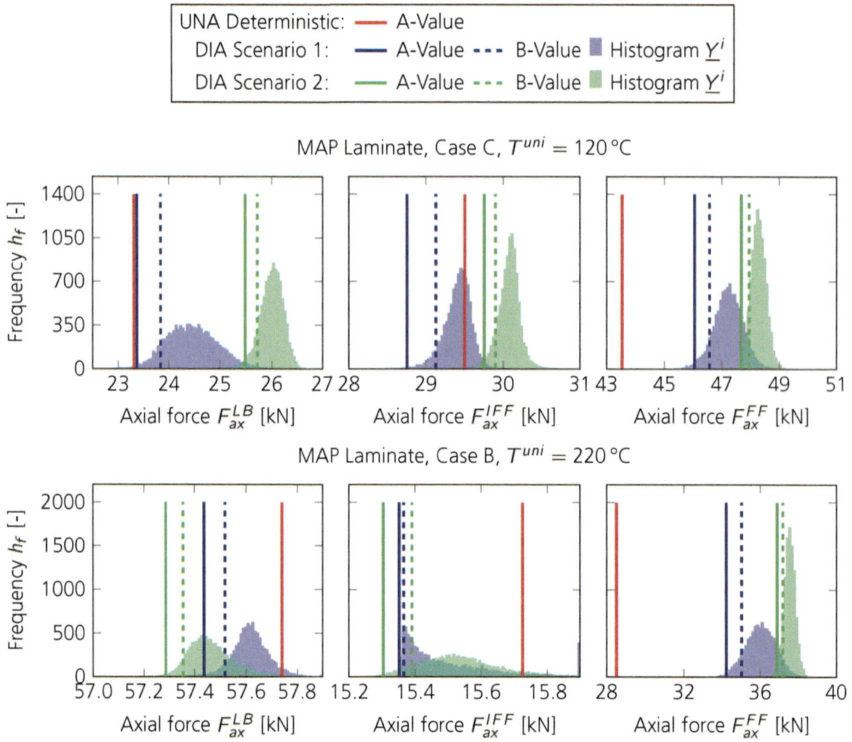

Fig. 6.9 Design values for the MAP Laminate. Comparison between UnA results (red) and DiA results where blue and green indicate statistical scenarios 1 and 2, respectively, and using solid and dashed lines to indicate A-values and B-values

of load-carrying capability are obtained for FF. For the MAP laminate, FF occurs in the thermally deformed state, which differs from the CP laminate where FF is initiated within the postbuckling domain. Failure is initiated at locations of highest compressive loading which in the current case is not at the center and is therefore distant from the highest thermal loading. However, the absolute values of the differences between UnA and DiA are similar to those found for the CP laminate, confirming the validity of these load-carrying potentials.

Conclusion on Surrogate Modelling and Probabilistic Analysis

The MAP example caused varying structural behavior for each of the three thermal cases. For thermal case A, inconsistent buckling behavior is observed. In this case, surrogate modelling and allowable determination is not conducted, since this violates the requirement of deformation consistency which is defined to enable a robust surrogate modelling procedure. For the other two cases, B and C, a consistent structural deformation behavior is observed throughout their parameter space. Hence, the procedure is continued in these cases and structural allowables are determined. The conclusions regarding model accuracy are similar as observed for the CP laminate.

The resulting potentials show slightly different values, which are resulting to the different buckling behavior that exists for the MAP laminate because of its orthotropic laminate setup. However, especially for FF, the potentials that exist by using the DiA procedure are noteworthy and significant at high temperatures.

6.5 Evaluation and Discussion

The research hypothesis of this work postulates that incorporating a detailed temperature distribution to structural analysis enables the identification and utilization of load-carrying capabilities compared to analysis procedures that assume uniform thermal conditions. To develop an DiA methodology that considers detailed temperature fields, working hypotheses are formulated to address the related challenges of temperature-dependent material modelling, design value evaluation, and thermal load uncertainties. Exemplary test studies are conducted in the previous sections, comparing the DiA procedure to the common UnA procedure.

At the beginning of this chapter, in Sect. 6.1, three general questions are defined to create the framework for these studies. Two different laminates are studied as test cases, the CP laminate and the MAP laminate. The knowledge gained for each structure is concluded at the end of the related sections. However, these conclusions are collected here to answer these questions and to comprehensively discuss the content of this work in reference to the working hypotheses and the research objective. The following sections are structured in accordance with the questions.

6.5.1 Structural Behavior

The main mechanisms that are considered by using a distributed temperature modelling compared to uniform thermal conditions contain:

- The lower average thermal conditions lead to a lower integral thermal strain caused along the preferential direction. This leads to varying stress development and can either be positive or negative for failure, as found for IFF and FF. In case of buckling, the lower average temperature leads to deformation states with lower compressive thermal stress and results in higher buckling loads.
- Residual stress contributes to the overall stress state and thus to failure. The magnitude of residual stress depends on the laminate setup and the local temperature. This contribution can especially be severe in transverse direction.
- Lower average temperature leads to a higher integral stiffness of the structure. This leads to higher buckling loads.
- The material strength components depend particularly on temperature. In distributed thermal conditions the highest stress is potentially caused in a certain

distance to the hotspot and thus in areas of higher strength. This leads to higher failure loads.
- Thermal gradients cause bending deformation, which can act as imperfection regarding buckling. This can lead to either positive or negative sensitivity to the buckling loads but includes a certain severity of preemptive buckling.

These effects, resulting to thermal loading, occur in superposition, making it difficult to derive specific recommendations. However, compared to uniform conditions, the present studies have shown that considering DiA leads to mostly positive potential for LB loads and FF loads. Negative potential (lower failure loads) is observed for IFF. On the other hand, the MAP laminate demonstrated an undesirable design case, where specific thermal load distributions act as imperfections and lead to preemptive buckling compared to other temperature distributions. Such behavior represent critical cases and are not accounted for in the UnA methodology. These potentials or the improved capability to identify critical cases can be achieved using the higher level of detail provided by DiA methodology.

These results provide insight into Working Hypothesis 3 (described on Sect. 2.2), which addresses the comparability between the structural behavior observed in uniform or distributed thermal conditions, respectively. In conclusion, Working Hypothesis 3 can only be partially confirmed, as there are critical cases observed through DiA which are not found in UnA conditions. The hypothesis further indicates load potentials when using DiA, which is confirmed for LB and FF loading scenarios. However limitations are observed for IFF and the possibility of preemptive buckling, as explained in detail above. Despite these restrictions, the ability to identify these critical cases is a positive outcome of the study.

6.5.2 Modelling and Analysis Procedure

The second question that is raised at the beginning of this chapter challenges the modelling procedures that are introduced using the DiA procedure. These involve additional modelling steps and conditions different to UnA and it is important to understand the benefits and limitations of each approach. The systematic approach considered in DiA can be divided into three modelling levels:

- Material modelling
- Finite Element Analysis and post processing
- Surrogate modelling

Material Modelling

Temperature dependent material property formulations are essential for analyzing structures under varying temperature conditions. The process of modelling the mechanical property relations is described in detail in Chap. 4, leading to allowable properties that account for both material and model uncertainties. This outcome mostly confirms Working Hypothesis 1, as discussed in Sect. 4.5. In contrast to that, the UnA methodology requires properties to be known only at a single temperature. It does not suffer from model errors since material properties can be applied directly. As noted in Sect. 4.5, some of the property models derived may be overly conservative due to outliers and poor statistical representativeness. In such cases, the model adds an unnecessary penalty to the properties, limiting the overall load carrying capability. High model uncertainties may lead to negative structural potentials when comparing the overall design strategies, if the margin of model uncertainty is higher than the potential gain from higher temperature resolution considering DiA. Therefore, minimizing the model errors is key to make the DiA procedure applicable and effective for exploiting load carrying capabilities. In this regard, it is crucial to have sufficient characterization data and to consider the approaches discussed in Sect. 4.5.

Finite Element Analysis

FEA is the second level of modelling used and is applied similarly in both UnA and DiA methodologies. The analysis setup, i.e. geometry, mesh, numerical parameters etc. are the same and the primary difference between the two is the varying thermal conditions applied in DiA, which leads to a higher level of detail and accuracy in the analysis, as explained in Sect. 6.5.1. Post-processing procedures to determine design values are similarly used in both methods. The higher level of detail and the accuracy provided by DiA results in greater confidence in the analysis, which is a positive outcome. However, this needs accurate knowledge about the thermal hotspot loading occurring in reality.

Surrogate Modelling

To address the challenge of thermal load uncertainties, surrogate models are used to approximate the structural behavior captured by FEA into functional relations. This procedure enables the efficient investigation of specified domains of thermal load distributions that are representative for the possible thermal load variation. This can serve for sensitivity analysis, determination of minimal values, or probabilistic analysis, as implemented in DiA and demonstrated within the scope of the studies described within this chapter. The applicability of this surrogate modelling procedure is ensured if the structural behavior is consistent along the investigated thermal parameter space. However, this is not the case for the thermal case A of the MAP laminate, which is therefore defined as an undesired design case and would not be further considered. Nevertheless, this methodology aligns with Working Hypothesis 4 and confirms that structural allowables can be determined to be representative of specific thermal load conditions, including specific worst cases.

Similar to material modelling, surrogate modelling introduces some model error to the analysis, which is not present in the UnA methodology. To assess the accuracy of the surrogate models, it is necessary to determine the associated error margins, which is part of the DiA process and is discussed throughout the evaluation of test cases. The errors found are generally small compared to the potential benefits of using DiA. However, if a margin of safety is desired, such as by accounting for these errors, part of the potential may be consumed. Therefore, it is important to further improve the surrogate modelling procedure. This includes gaining more experience on how to efficiently define the thermal parameter space that allows the model to be accurate or considering more suitable surrogate model approaches. These definitions depend heavily on the specific use case being investigated.

Conclusion
To ensure a robust design, it is essential to consider the model errors introduced by the DiA procedure. However, these errors can act as a penalty on the overall performance when comparing the pure analysis strategy of DiA to the deterministic UnA approach. To verify, if the DiA procedure is beneficial, these errors and potentials can be added. However, since the DiA methodology is a systematic approach, there is scope for further improvement and gaining experience in modelling techniques that can help minimize these errors. Therefore, while it is important to acknowledge and consider the model errors, it is equally essential to continue advancing the methodology to reduce these errors and make it more efficient and effective.

6.5.3 Potentials and Transferability

The third question that is introduced at the beginning of this chapter addresses the resulting potentials in load-carrying capability that can be achieved using DiA.

Section 6.5.1 explains the structural mechanisms that are considered by DiA and that produce a more detailed and accurate representation of the structural behavior. These mechanisms finally cause the potential in load carrying capability. However, the magnitude of the resulting load potential is highly dependent on the specific structural setup and boundary conditions used. It further varies between the design values that are considered. Therefore, it is not possible to estimate a general margin of additional load-carrying capability that can be utilized by DiA compared to UnA. Two factors that scale the resulting margins are the boundary conditions used and the thermal gradient within the structure. More relaxed boundary conditions result in less potential, since thermal deformation and stress can relax. However, it is important to ensure that the analysis accurately represents the real-world environment, since also adjacent structures could be thermally loaded leading to a global stress state. Additionally, the thermal gradients that exist within the structure scales the resulting potential and the average temperature in the structure correlates to the effective properties and the integral thermal load. Further, thermal gradients can causes thermal deformation, which can become sensitive and severe.

To ensure a robust design, it is essential to consider the model errors introduced by the DiA procedure. These errors are explained in Sect. 6.5.2 and can act as a penalty on the overall performance when comparing the pure design strategy to the common UnA approach. To verify, if the DiA procedure is benefitial, these errors and potentials can be added. However, since the DiA methodology is a systematic approach, there is scope for further improvement and gaining experience in modelling techniques that can help minimize these errors. Therefore, while it is important to acknowledge and consider the model errors, it is equally essential to continue advancing the methodology to reduce these errors and make it more efficient and effective.

6.5.4 Design Value Determination

Working Hypothesis 2 aims to investigate the utilization of design criteria under distributed thermal conditions and to verify their applicability within the scope of the DiA procedure. The working hypothesis can be confirmed, but it requires specific explanations. The methods implemented and used for evaluating the design criteria, namely LB, IFF, and FF, are robust and worked in all cases. However, the results require careful consideration and have some limitations.

Buckling
The methodology used to determine buckling is generally robust and worked in all cases analyzed. However, it should be noted that different buckling phenomena can occur depending on the structural setup, and these variations can naturally lead to different results in buckling load. Therefore, a careful study of the structural behavior that occurs during buckling is still needed to fully understand and interpret the data. Recording additional physical relations, e.g. strain behavior, that complement the structural stiffness progression could help to identify and classify different buckling phenomena, enabling engineers to better understand the sensitivity of the buckling behavior.

Failure
The strength criteria applied within the analyses conducted within this chapter mainly focuses on the Puck criteria for evaluating the IFF and FF behavior. However, other failure criteria such as MaxStress, Hashin, and Cuntze are also utilized in parallel. The detailed equations for these criteria are provided in Appendix Chap. B, and their application is discussed in Sect. 5.3.1. While different criteria can lead to slightly different design values, the occurrence of such discrepancies is rare. The deviations are small if they occur (results vary by a single increment of FEA). This is mainly caused by the dominating stress states rather than the criteria themselves. Overall, the results obtained from different criteria are consistent, indicating their effectiveness in assessing the failure behavior of composite structures.

6.6 Summary and Conclusion

This chapter applies the analysis procedures and material models developed throughout this work, designated as DiA procedure, to analyze a representative skin-field structure. These procedures apply detailed temperature distributions within structural analysis and are an advancement compared to the present methodology applying uniform thermal conditions, abbreviated here as UnA. The investigation applies two different laminate stackings, namely a cross-ply (CP) setup with equal effective properties in x and y direction and a modified angle-ply (MAP) setup with different effective properties in x and y directions. For each laminate setup, three thermal cases are studied with different temperature margins: a low thermal load case, a common thermal load case at the maximum allowable service temperature of the material, and a high-temperature case with a service temperature equal to the critical temperature of the material (PES T_g), which is beyond common design approaches. Each of these variations is used to compare DiA and UnA methodologies and their results. This serves for studying the research objective.

Distributed thermal conditions cause various structural effects that act in superposition. In uniform conditions, these effects are not present or in different extent and may lead to varying sensitivities to the design values. Using the DiA procedure, the uncertainty of thermal load variations is considered by the resulting structural allowables, which are design values that are similar to common design loads but representative for a specified thermal load domain. Compared to the analysis procedure using uniform conditions, potentials in the structural load carrying capability are observed for LB and FF. These depend on the structural setup and the magnitude of thermal load applied. In addition, negative potentials, i.e. lower design values, are observed for IFF. Although this mismatch is relatively low, it means that the uniform analysis approach is not always conservative. Further, it could be observed, that some specific thermal loads can act as critical imperfection causing premature buckling. Such phenomenology is similarly not covered by uniform temperature conditions. This is a critical and relevant observation.

In conclusion, the results confirm the research objective. However, the observations found have to be differentiated. Load carrying potentials exist and can be utilized using the developed DiA procedure for local buckling and fiber failure clearly confirming the hypothesis. In case of inter-fiber failure, mostly lower design values are captured by the new procedure. This and also the finding of possible premature buckling, however, contribute to confirm the hypothesis, since they improve the confidence into analysis. In this regard, the consideration of a higher level of detail through the applied temperature distribution serves for more robust designs and finally for more efficient designs.

References

1. M. Liebisch et al., Robustness analysis of CFRP structures under thermomechanical loading including manufacturing defects. Athens (2019). https://www.sucohs-project.eu/publications-presentations
2. M. Liebisch, T. Wille, Exploitation of high performance structures at elevated temperatures. Aerospace Europe Conference 2020 (2020)

Chapter 7
Summary

7.1 Summary and Conclusion

This work addresses the structural analysis of lightweight CFRP composite structures that are exposed to thermomechanical load conditions and specifically refers to applications where thermal loads are caused by external heat sources. Such conditions are particularly challenging for CFRP materials, as their behavior is orthotropic and highly dependent on temperature. Thermal expansion can significantly contribute to the deformation state and stress development. However, typical structural sizing makes use of simplified uniform thermal conditions and thermal worst-case scenarios. They do not accurately reflect realistic scenarios but are supposed to be conservative. In reality, structures exposed to external heat sources develop temperature distributions. They exhibit a hotspot with temperature magnitude equivalent to the worst-case temperature, and develop lower temperatures in distance to the hotspot. This work hypothesizes that the deterministic analysis procedure using uniform thermal conditions is overly conservative, and that load-carrying potentials can be better exploited using a detailed consideration of the spatial temperature distribution in the structure.

In uniform conditions, structural analysis can be performed with material properties characterized at a single temperature. However, in distributed thermal conditions, properties must be known as a function of temperature to accurately analyze the mechanical structural behavior in distance to the hotspot. To this end, an extensive material characterization program is conducted to study the behavior of a high-temperature resistant CFRP material and to provide suitable data for subsequent material modelling. Within this work, a toughened cyanate ester matrix material was used, resulting in an intermediate transition that caused a drop in material properties at around 225 °C. The characterization program covered temperatures from room temperature up to 250 °C and partially above and the test procedures used are typical industrial standards. All material properties, except for transverse tensile strength

M. Liebisch, *Design Load Allowables for Composite Plates Exposed to Thermomechanical Loads*, Mechanics and Adaptronics,
https://doi.org/10.1007/978-3-031-71352-1_7

and compression strength in the fiber direction, were adequately captured. A linear elastic behavior was observed for all properties and temperatures, except for shear. The captured uncertainties do not refer to a suitable statistical database but provide similar ranges compared to values from the literature. The material characterization provides suitable data for modelling and provides valuable insights into the behavior of CFRP materials over a range of temperatures. The findings are further useful for guidance in similar experimental studies in the future.

Building on the material characterization results, material modeling was conducted to formulate material properties as a function of temperature. To achieve this, a model approach proposed by Mahieux is employed, which is based on the phenomenological behavior of polymers over temperature. In this approach, polymer bond breakage correlates with temperature, ultimately leading to macroscopic property degradation. This approach is selected as it provides a useful relationship between temperature and property behavior, particularly when the dataset has a low number of experimental points. DMA measurements were used to determine the model parameters. The final property models use these parameters and define the individual property magnitudes based on the static material characterization results. This procedure allowed to account for the individual elastic and strength properties, including their specific behavior over temperature. Additionally, the model was enhanced to consider both material uncertainty and model uncertainty. Model uncertainty was particularly important because material allowables are required to be conservative. Through comparing the resulting models to the characterization results, conservative property margins are found for all properties. In most cases, the errors between model and measurement were lower than the material uncertainty, indicating good accuracy. However, some exceptions caused higher model errors. In these cases, the model error could be attributed to the transition area, where properties significantly drop in magnitude, and was caused mainly by poor statistical representativeness of the database.

Chapter 5 describes the procedure used for structural analysis under thermomechanical load conditions with distributed temperature, designated as DiA procedure. A structural analysis involves multiple sequential FEA to determine the resulting temperature, resulting thermal deformation, and the deformation that results from axial compression loading. The analysis also considers residual stress, which finally superposes with thermally induced stress and is especially sensitive to material failure. Various methods for evaluating design criteria are discussed and implemented for local buckling and material strength (IFF, and FF). Each structural analysis is performed for a specific temperature distribution and captures the design loads. The relation of thermal parameters, describing the temperature distribution, and design loads defines the training points. Various of these structural analyses are performed with varying the thermal conditions and obtain a set of such training points. Based on that surrogate modelling is done to establish a functional relationship between the design values and variations in the temperature distribution. These surrogate models are subsequently used to perform probabilistic analysis and to derive structural allowables. These allowables represent the equivalent to the typical deterministic analysis values analyzed for uniform conditions, but they are representative for a specified range of thermal conditions. This approach serves to investigate the research hypothesis.

The final chapter of this study demonstrates the analysis procedures developed throughout the work. The chapter presents a comparison of the analysis procedure using distributed temperature (DiA) with common analyses that apply uniform conditions (UnA) for two different skin-field like structures having different laminate stackings. The first example is a cross-ply (CP) setup with equal properties in both directions, and the second one is a modified angle-ply setup (MAP) with different properties in longitudinal and transversal directions. Three different thermal test cases with varying temperature magnitude are investigated for both laminates. These examples provide a comprehensive overview of the sensitivities that arise from distributed thermal conditions.

Several structural mechanical effects are observed acting in superposition. The DiA procedure is utilized to determine structural allowables which are representative for the distributed conditions. Compared to the uniform approach, they obtain load-carrying potentials for LB and FF. However, deficiencies of the uniform analysis procedure are observed for IFF, where distributed conditions can lead to lower design values. An undesirable case is generated where specific thermal load cases produce preemptive buckling, which is not observed in the uniform analysis. In conclusion, the higher level of detail and the consideration of thermal load variations by using the new method offer several advantages and confirm the research objective. The procedure leads to more efficient structures, indicated by higher design values, or improves the understanding of the structural behavior in severe cases and thus supports more robust designs.

7.2 Outlook

The DiA methodology developed throughout this work is a systematic approach, that can be utilized and enhanced in its individual procedure items. Further need for research is identified.

Material characterization and modelling have proven to be effective in determining temperature-dependent allowables for structural design. However, one limitation of current research is the low number of samples and their limited statistical representativeness, which introduces uncertainty into the results and requires unnecessary high margins of conservativity. Future research could include more extensive characterization programs to capture more experiences on this. Future studies should also focus on developing more efficient ways to quantify and account for uncertainties. Opportunities for this are mentioned in the work and should be pursued to fully exploit the potential of the material.

In addition, future research could aim to enhance the material models to account for plasticity. As demonstrated in this work, the onset of polymer yield stress decreases with increasing temperature. First studies to incorporate this detail into the composite models is available in [1]. Such enhancement could help to better predict material behavior under high thermomechanical loading conditions, to fully exploit its potentials and to further optimize structural concepts.

Suitable methods for evaluating design values are crucial for accurate structural analysis and assessment. The implemented buckling criterion has proven effective, but engineers must still carefully investigate buckling sensitivities to ensure reliable results and robust design. This study highlights the importance of such investigations, particularly in the context of thermal buckling and deformation. Developing a methodology to rate the criticality of deformation states and associated buckling phenomena would be valuable in supporting these investigations. This could, for instance, include a strain-based criterion that is used in conjunction with the currently implemented approach.

Furthermore, verifying failure theories at elevated temperatures up to the glass transition temperature under multiaxial loading states is an important and open research topic for applications in thermomechanical load conditions. Initial studies in this direction could be combined with more comprehensive property characterization studies to better understand material behavior under these conditions.

An important future aspect is to apply the DiA procedure for structural analyses of complex structures, e.g. stiffened panels, to investigate related potentials and the sensitivity of boundary conditions. This investigation could also address the implementation of more appropriate surrogate modelling strategies or improve error assessment. Moreover, the DiA methodology can be utilized in many directions, and for instance applied for the purpose of tailored designs. By using DiA, structures could be designed to reliably cause hotspots and experience the highest mechanical loading in specific and distant areas, leading to higher failure loads. The methodology can also increase confidence in analysis, leading to a reduction in thermal margins of safety and allowing for service temperatures closer to the glass transition temperature, finally resulting in more efficient and lightweight structures. A careful study of the variety of topics and conservatism related to thermal degradation and deterioration as well as experimental proof are recommended.

Reference

1. K.L. Reifsnider, V. Tamuzs, S. Ogihara. On nonlinear behavior in brittle heterogeneous materials. Compos. Sci. Technol. **66**(14), 2473–2478 (2006). ISSN: 02663538. https://doi.org/10.1016/j.compscitech.2006.04.003

Appendix A
Material Characterization

A.1 Overview to the Characterization Program

Table A.1 provides an overview to the standards used and Fig. A.1 shows a schematic drawing of the test fixtures used.

Table A.1 Test standards used for characterization of mechanical material properties. Specimen dimensions are defined by length × width × thickness (L × W × T). F denotes the free length

Property	Details
Tension	• ASTM D3039 [1.] • Single side strain gauges used • No tabs used • Specimen dimension fiber direction: 250 mm · 15 mm · 0.72 mm, F = 140 mm • Specimen dimension transverse direction: 175 mm · 25 mm · 2.4 mm, F = 125 mm
Compression	• EN 2850 [2] • Specimen dimension: 80 mm · 12.5 mm · 2.4 mm • Different test setups for strength and modulus determination • No tabs but back-to-back strain gauges used for modulus determination, F = 12.7 mm • Tabs of 2mm thickness used for strength determination, F = 5 mm
Shear	• In-plane shear testing according to ASTM D3518 [3] • The standard is in agreement with ASTM D3039 but uses ±45° stacking • Specimen dimension: 250 mm · 25 mm · 1.96 mm, F = 100 mm • Shear strain to be determined from perpendicular strain gauges
Interlaminar shear	• ILSS according to DIN EN 2563 [4] • Specimen dimension: 20 mm · 10 mm · 2 mm • Three point bending test, no tabs, no strain measurement

M. Liebisch, *Design Load Allowables for Composite Plates Exposed to Thermomechanical Loads*, Mechanics and Adaptronics,
https://doi.org/10.1007/978-3-031-71352-1

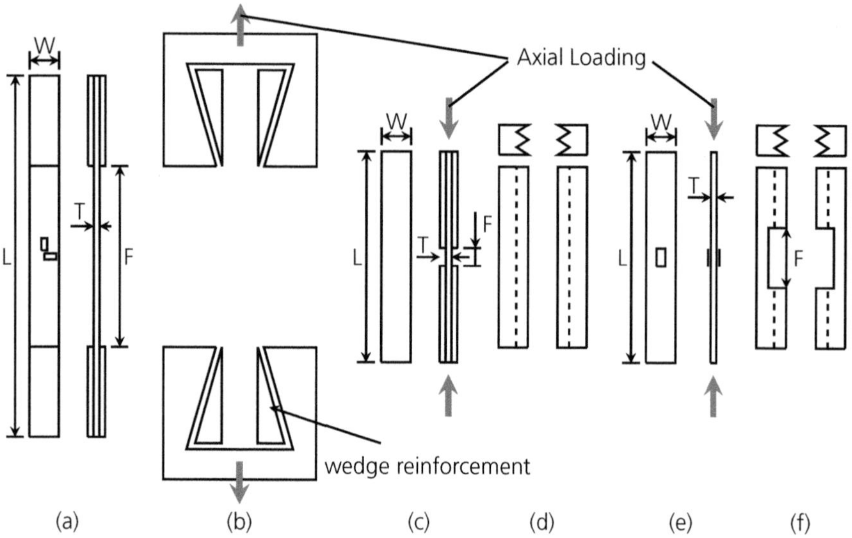

Fig. A.1 Schematic overview to test fixtures and specimen dimensions used within material characterization. In line with Table A.1. **a** Specimen dimension within tensile testing, dimensions vary between fiber direction, transverse direction and shear testing. **b** Schematic drawing of the wedge clamping device applied within tensile testing. Compression specimen and related clamping devices according to EN 2850 for strength determination (Type B1) in (**c**) and (**d**) and modulus determination (Type B2) in (**e**) and (**f**)

A.2 Tensile Testing

Fiber Direction

Tables A.2 and A.3 present an overview to the result of tensile testing in fiber direction. Figures A.2–A.6 show the measurement curves these results are derived from.

Table A.2 Overview to characterization results of tension modulus in fiber direction

Sign	Unit	23 °C	150 °C	200 °C	250 °C	275 °C
n	–	2	3	3	4	1
μ	MPa	162274.4	167110.4	162780.0	165789.1	n.a.
$\tilde{\mu}_{RT}$	%	100.0	103.0	100.3	102.2	n.a.
s	MPa	4239.9	8753.6	4227.4	7760.5	n.a.
cv	%	2.6	5.2	2.6	4.7	n.a.

Table A.3 Overview to characterization results of tension strength in fiber direction

Sign	Unit	23 °C	150 °C	200 °C	250 °C	275 °C
n	–	3	3	3	5	1
μ	MPa	2712.5	2236.9	2371.9	2338.3	n.a.
$\tilde{\mu}_{RT}$	%	100.0	82.5	87.4	86.2	n.a.
s	MPa	362.8	308.1	112.3	390.4	n.a.
cv	%	13.4	13.8	4.7	16.7	n.a.

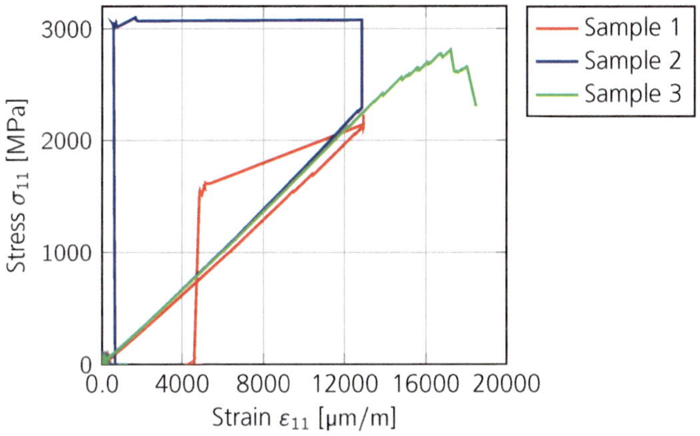

Fig. A.2 Tension characterization in fiber direction at 23 °C

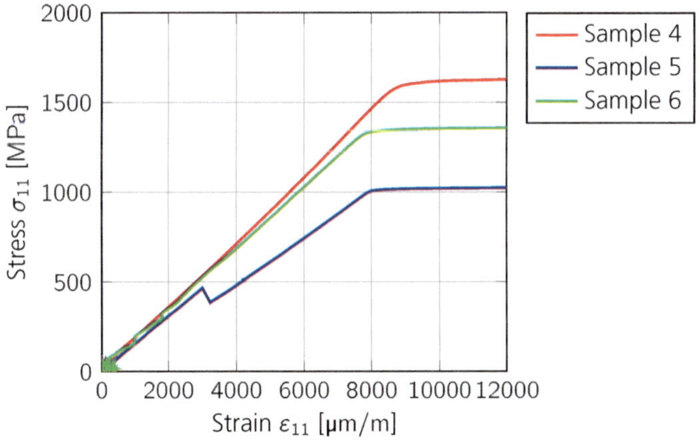

Fig. A.3 Tension characterization in fiber direction at 150 °C

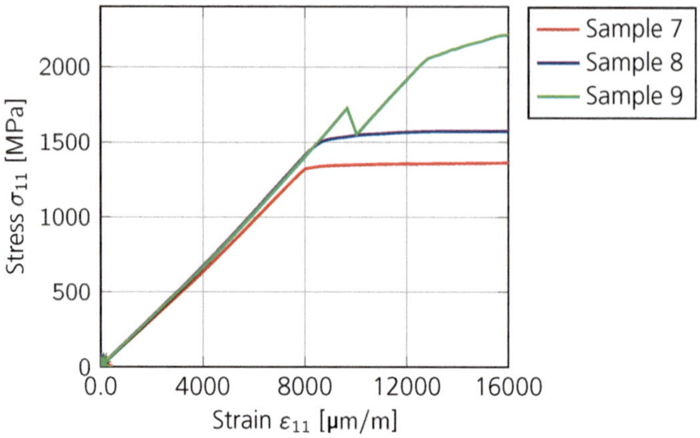

Fig. A.4 Tension characterization in fiber direction at 200 °C

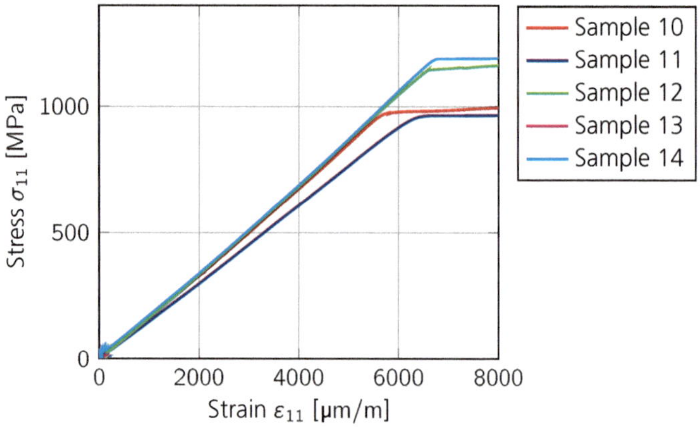

Fig. A.5 Tension characterization in fiber direction at 250 °C

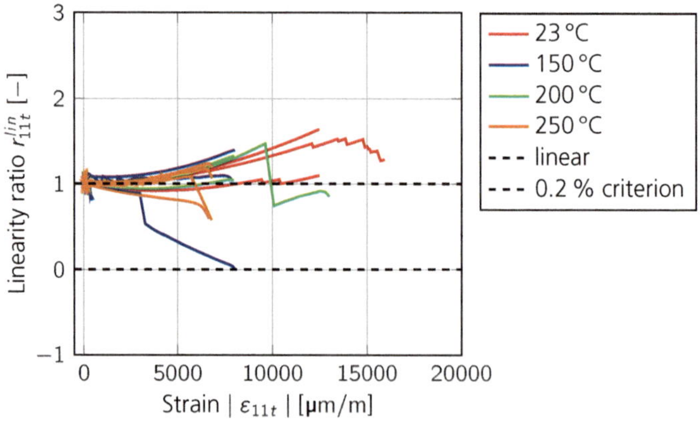

Fig. A.6 Linearity evaluation of the elastic behavior under tensile loads in fiber direction

Transverse Direction

Tables A.4 and A.5 present an overview to the result of tensile testing in transverse direction. Figures A.8–A.13 show the measurement curves these results are derived from.

Table A.4 Overview to characterization results of transverse tension modulus

Sign	Unit	23 °C	150 °C	200 °C	225 °C	250 °C	275 °C
n	–	2	2	2	2	2	2
μ	MPa	9524.1	8673.9	7728.3	6007.8	5469.6	4677.2
$\tilde{\mu}_{RT}$	%	100.0	91.1	81.1	63.1	57.4	49.1
s	MPa	322.8	360.4	14.4	584.4	279.9	26.0
cv	%	3.4	4.2	0.2	9.7	5.1	0.6

Table A.5 Overview to characterization results of transverse tension strength

Sign	Unit	23 °C	150 °C	200 °C	225 °C	250 °C	275 °C
n	–	3	3	3	3	2	2
μ	MPa	14.6	11.5	10.8	9.5	11.2	10.3
$\tilde{\mu}_{RT}$	%	100.0	78.8	73.7	65.1	76.4	70.7
s	MPa	4.2	2.4	1.7	2.5	0.4	0.0
cv	%	28.9	20.7	16.1	26.0	3.6.	0.4

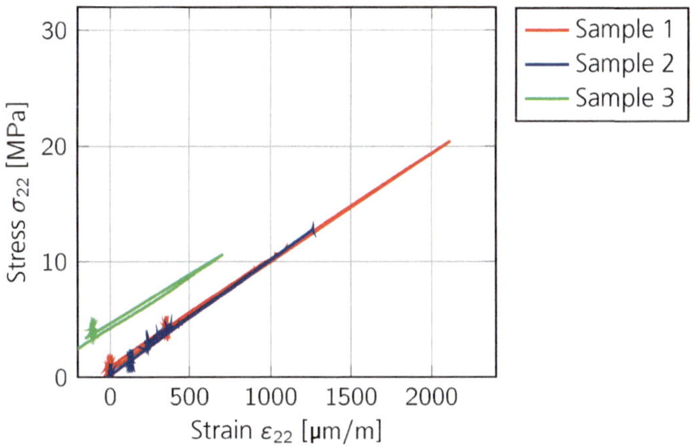

Fig. A.7 Tension characterization in transverse direction at 23 °C

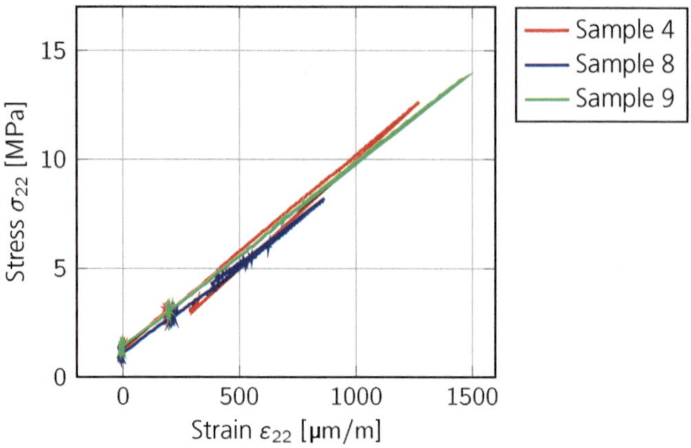

Fig. A.8 Tension characterization in transverse direction at 150 °C

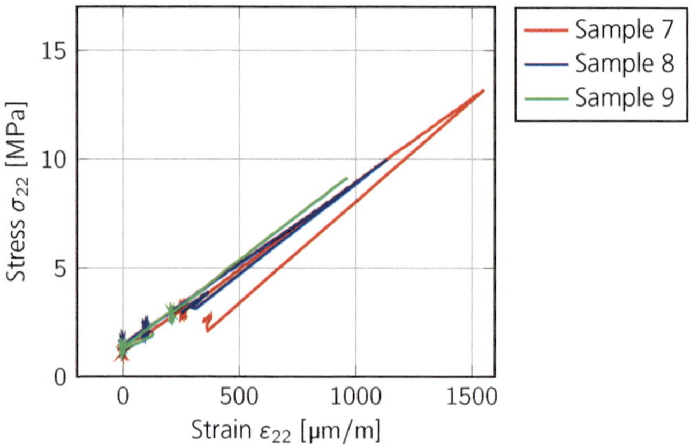

Fig. A.9 Tension characterization in transverse direction at 200 °C

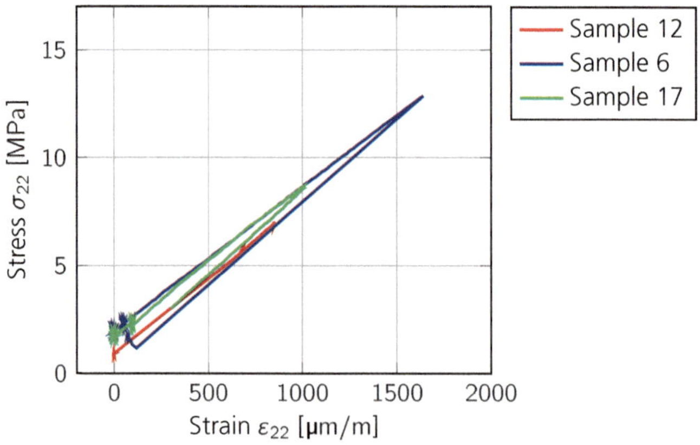

Fig. A.10 Tension characterization in transverse direction at 225 °C

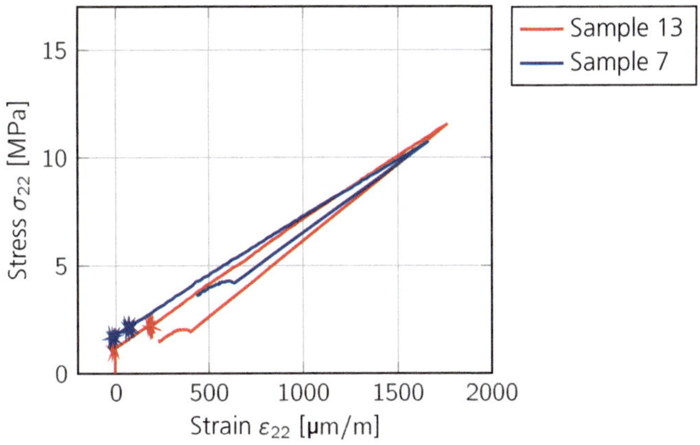

Fig. A.11 Tension characterization in transverse direction at 250 °C

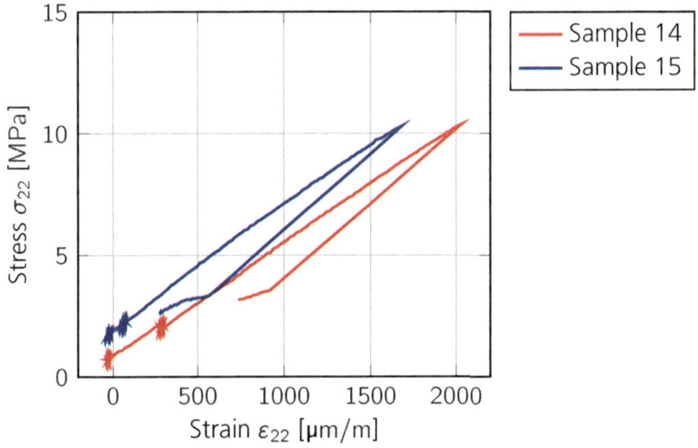

Fig. A.12 Tension characterization in transverse direction at 275 °C

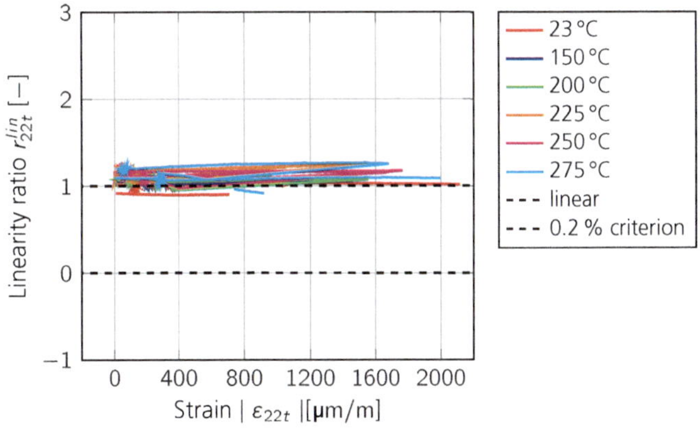

Fig. A.13 Linearity evaluation of the elastic behavior under tensile loads in transverse direction

A.3 Compressive Testing

Fiber Direction

Tables A.6 and A.7 present an overview to the result of compressive testing in fiber direction. Figures A.14–A.23 show the measurement curves these results are derived from.

Table A.6 Overview to characterization results of fiber direction compression modulus

Sign	Unit	23 °C	150 °C	200 °C	225 °C	250 °C
n	–	4	3	2	2	3
μ	MPa	143570.5	139372.7	146754.4	149843.5	166834.8
$\tilde{\mu}_{RT}$	%	100.0	97.1	102.2	104.4	116.2
s	MPa	7079.3	3658.1	1082.3	8845.1	11013.9
cv	%	4.9	2.6	0.7	5.9	6.6

Table A.7 Overview to characterization results of fiber direction compression strength

Sign	Unit	23 °C	150 °C	175 °C	200 °C	225 °C	237 °C	250 °C	275 °C	300 °C
n	–	6	3	3	3	3	3	3	3	3
μ	MPa	964.6	837.6	719.3	800.1	592.3	371.6	190.8	114.2	96.6
$\tilde{\mu}_{RT}$	%	100.0	86.8	74.6	82.9	61.4	38.5	19.8	11.8	10.0
s	MPa	92.6	39.2	35.7	47.5	26.0	88.9	24.6	3.3	6.2
cv	%	9.6	4.7	5.0	5.9	4.4	23.9	12.9	2.9	6.4

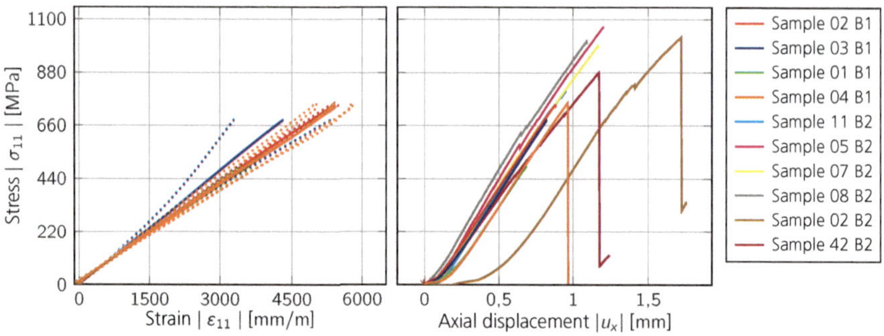

Fig. A.14 Compression characterization in fiber direction at 23 °C

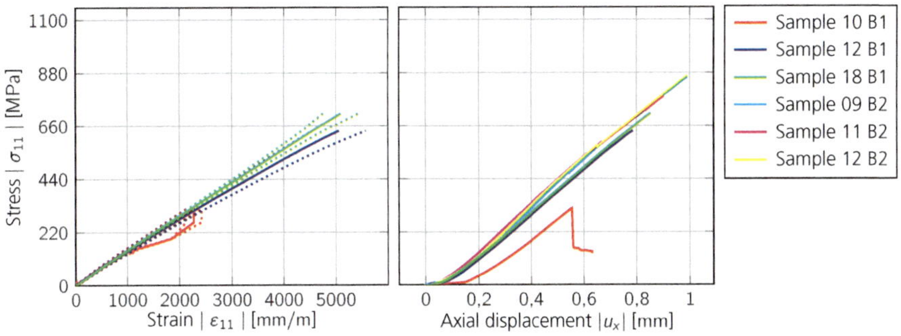

Fig. A.15 Compression characterization in fiber direction at 150 °C

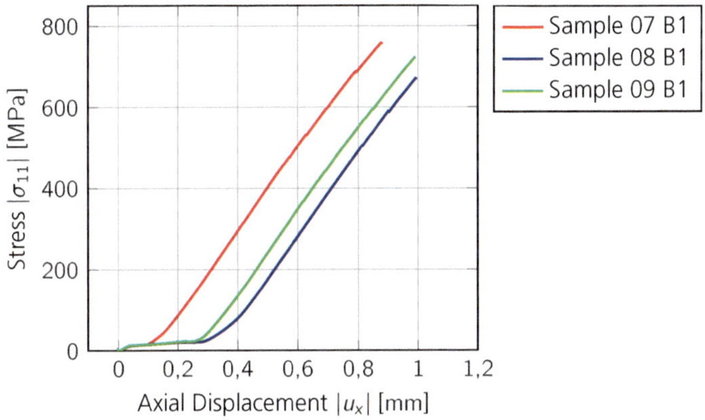

Fig. A.16 Compression characterization in fiber direction at 175 °C

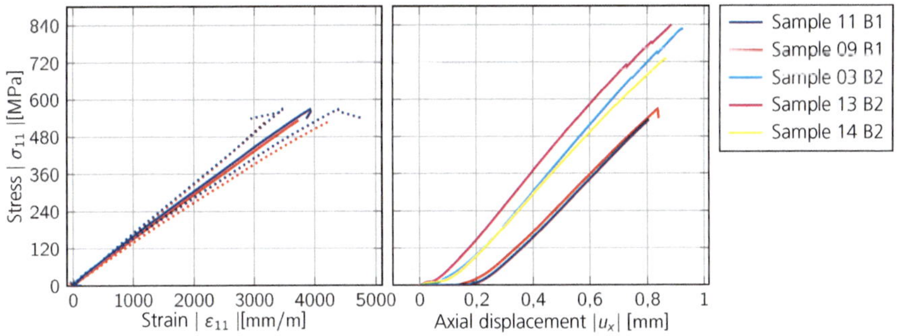

Fig. A.17 Compression characterization in fiber direction at 200 °C

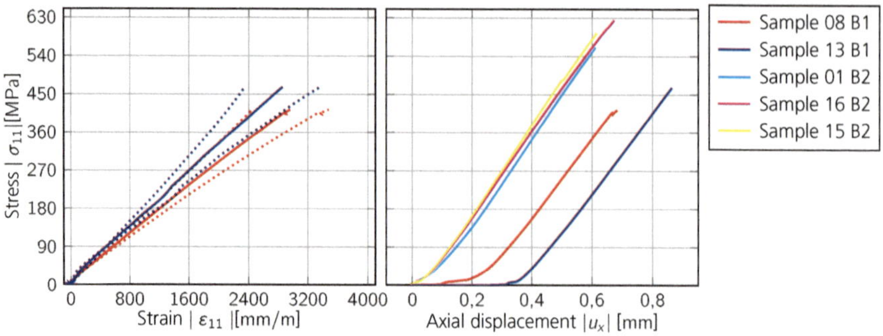

Fig. A.18 Compression characterization in fiber direction at 225 °C

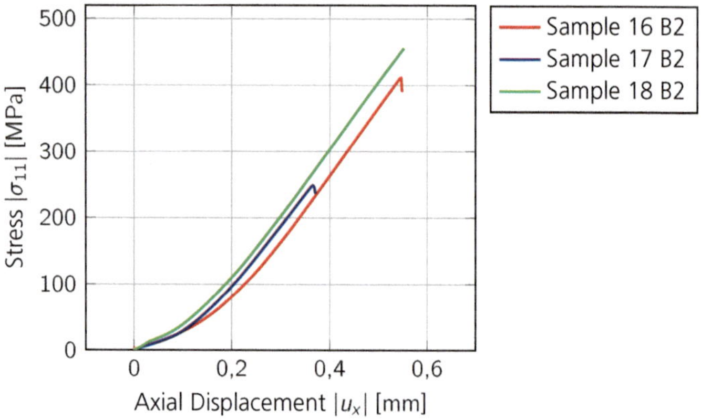

Fig. A.19 Compression characterization in fiber direction at 237 °C

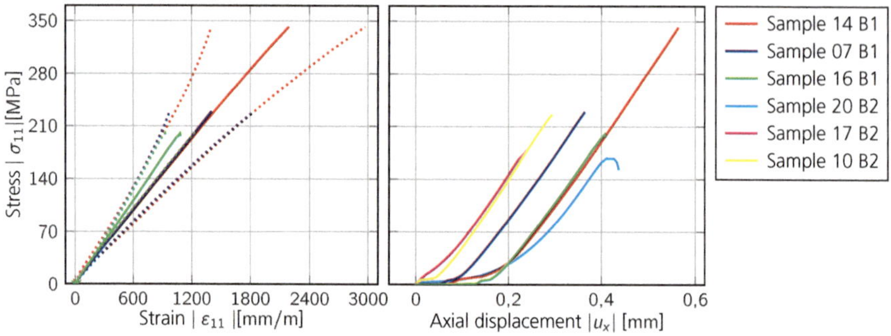

Fig. A.20 Compression characterization in fiber direction at 250 °C

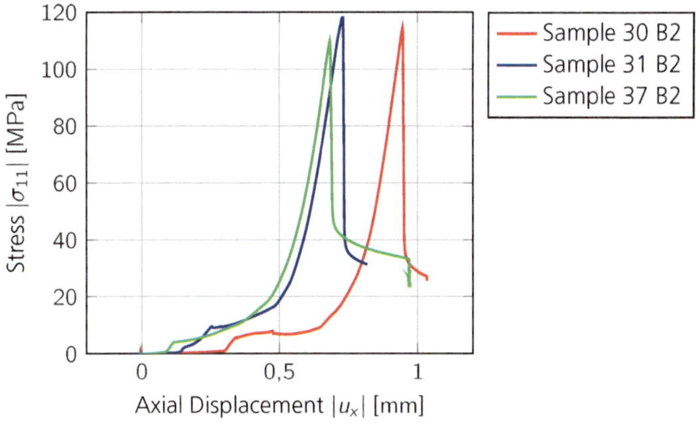

Fig. A.21 Compression characterization in fiber direction at 275 °C

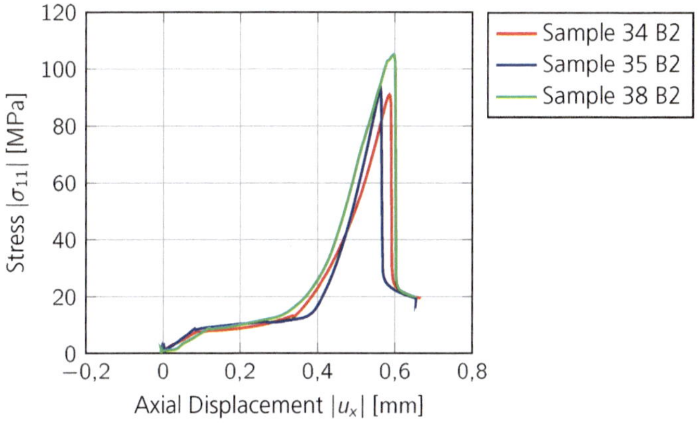

Fig. A.22 Compression characterization in fiber direction at 300 °C

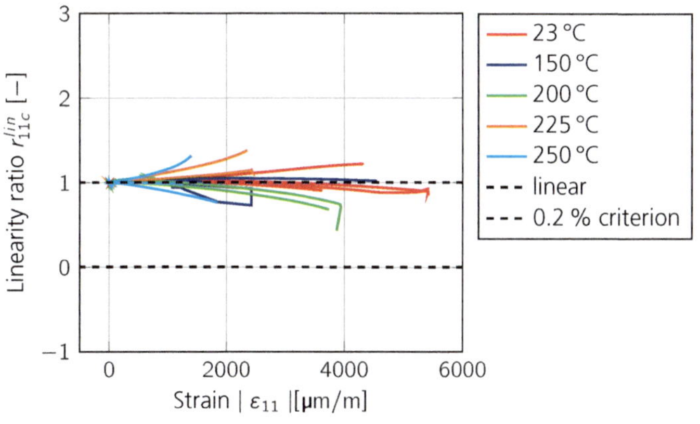

Fig. A.23 Linearity evaluation of the elastic behavior under compressive loads in fiber direction

Transverse Direction

Tables A.8 and A.9 present an overview to the result of compressive testing in transverse direction. Figures A.24–A.33 show the measurement curves these results are derived from.

Table A.8 Overview to characterization results of transverse compression modulus

Sign	Unit	23 °C	150 °C	200 °C	225 °C	250 °C
n	–	5	2	3	2	2
μ	MPa	9700.8	7129.3	6065.2	5413.9	4509.8
$\tilde{\mu}_{RT}$	%	100.0	73.5	62.5	55.8	46.5
s	MPa	280.7	53.5	441.5	163.1	365.6
cv	%	2.9	0.8	7.3	3.0	8.1

Table A.9 Overview to characterization results of transverse compression strength

Sign	Unit	23 °C	100 °C	150 °C	175 °C	200 °C	225 °C	250 °C	275 °C	300 °C
n	–	4	3	3	3	3	4	3	3	3
μ	MPa	236.3	212.2	168.6	136.8	136.6	92.8	74.9	25.6	19.0
$\tilde{\mu}_{RT}$	%	100.0	89.8	71.3	57.9	57.8	39.3	31.7	10.9	8.0
s	MPa	11.9	3.7	5.7	5.7	8.0	4.8	17.5	2.2	0.8
cv	%	5.0	1.8	3.4	4.2	5.9	5.2	23.4	9.8	4.2

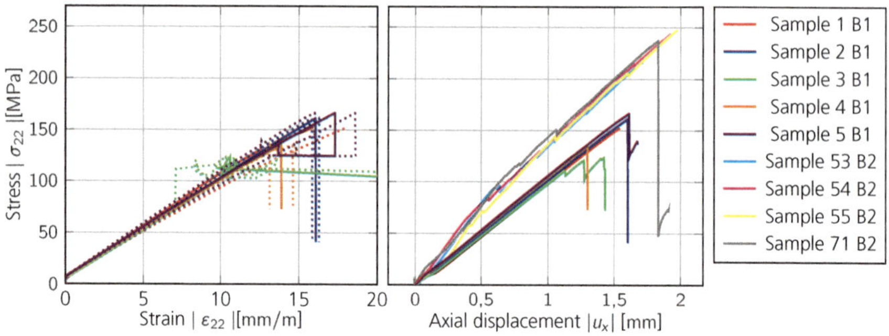

Fig. A.24 Compression characterization in transverse direction at 23 °C

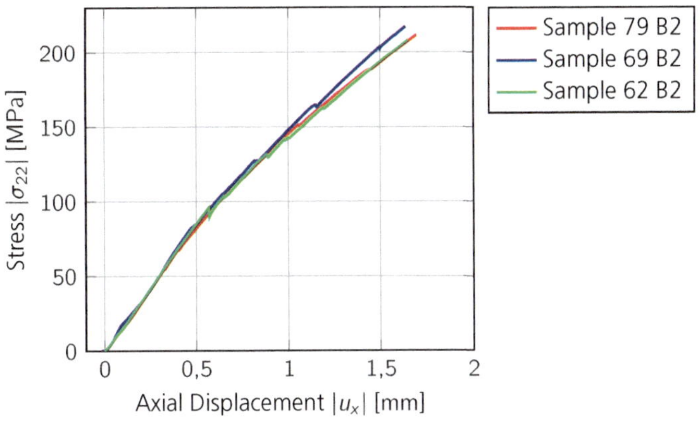

Fig. A.25 Compression characterization in transverse direction at 100 °C

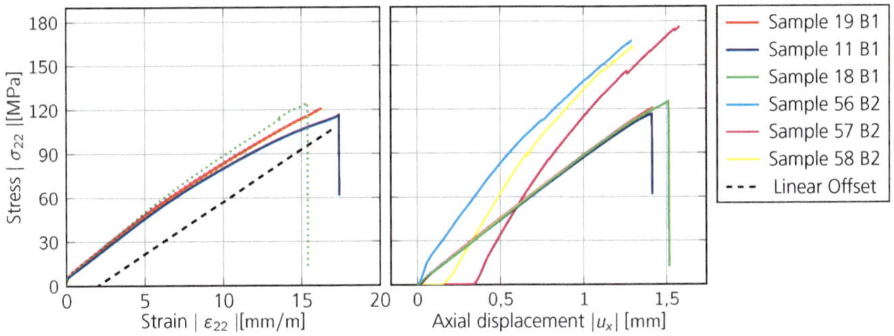

Fig. A.26 Compression characterization in transverse direction at 150 °C

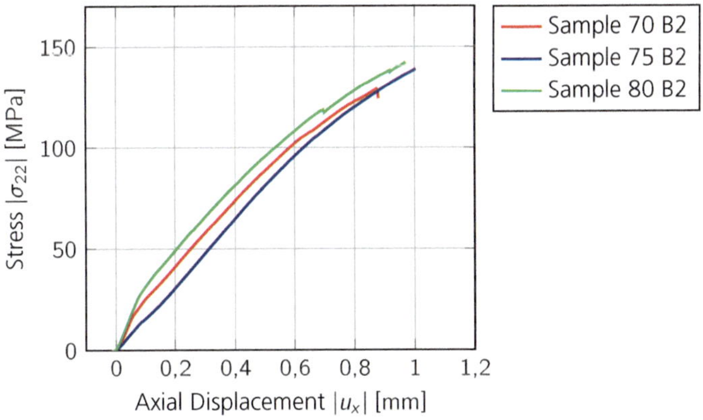

Fig. A.27 Compression characterization in transverse direction at 175 °C

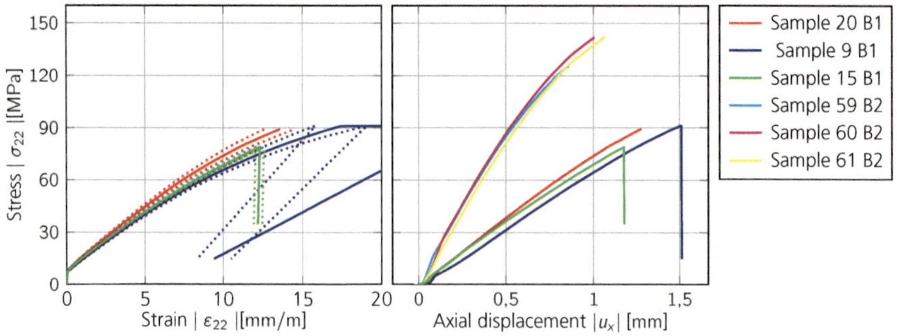

Fig. A.28 Compression characterization in transverse direction at 200 °C

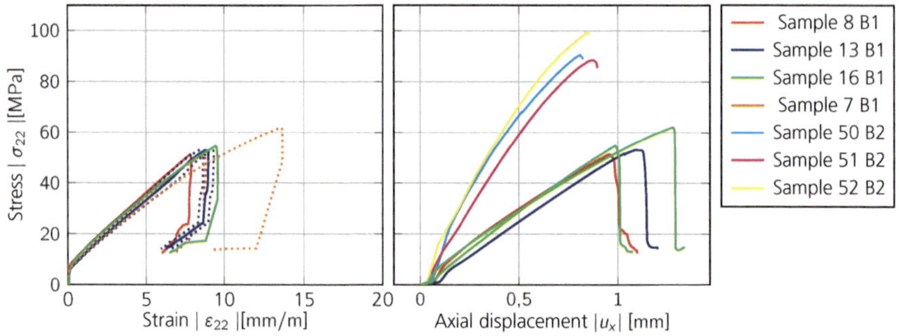

Fig. A.29 Compression characterization in transverse direction at 225 °C

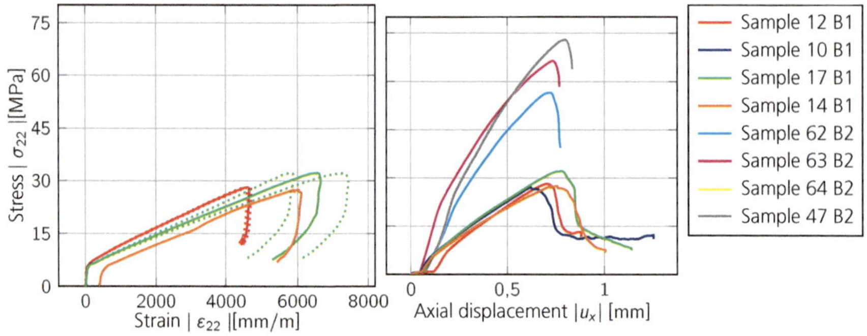

Fig. A.30 Compression characterization in transverse direction at 250 °C

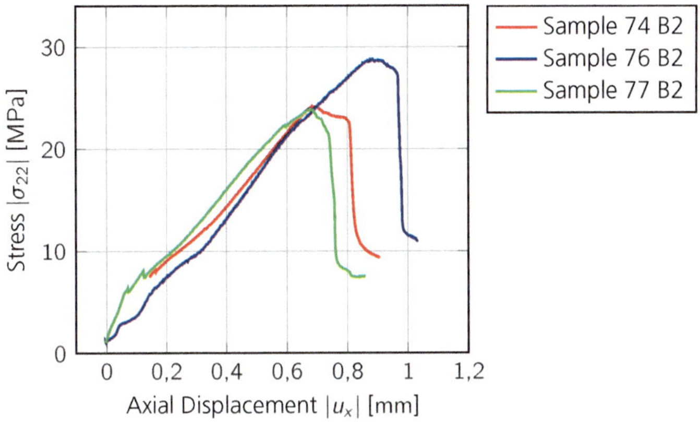

Fig. A.31 Compression characterization in transverse direction at 275 °C

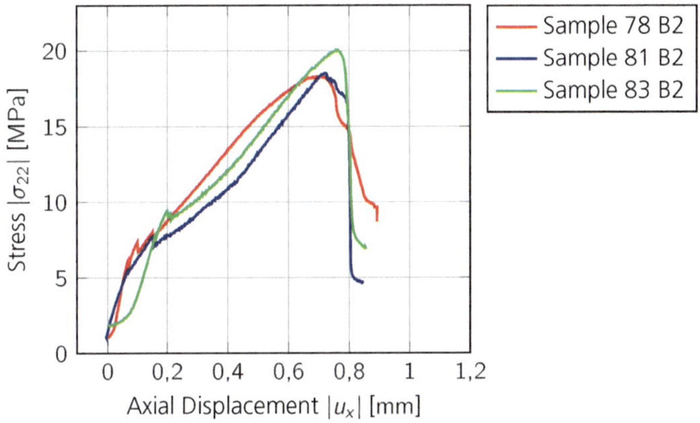

Fig. A.32 Compression characterization in transverse direction at 300 °C

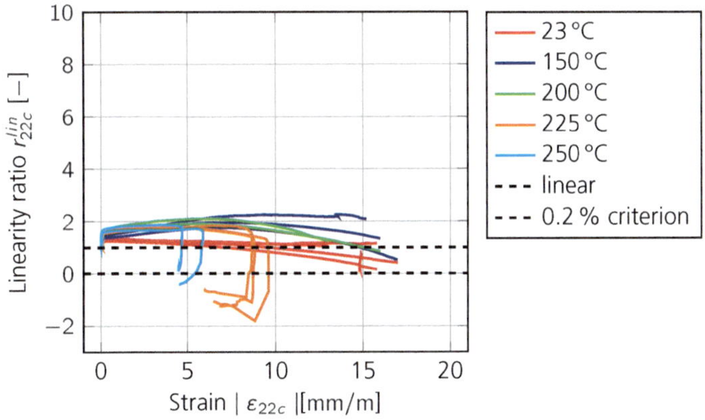

Fig. A.33 Linearity evaluation of the elastic behavior under compressive loads in transverse direction

A.4 In-Plane Shear Testing

Tables A.10 and A.11 present an overview to the result of IPS testing. Figures A.34–A.39 show the measurement curves these results are derived from.

Table A.10 Overview to characterization results of the in-plane Shear modulus

	Sign	Unit	RT	150 °C	200 °C	225 °C	250 °C
Shear modulus G_{12}	n	–	4	3	3	3	3
	μ	MPa	5391	3012	1767	1245	734
	$\tilde{\mu}_{RT}$	%	100.0	55.9	32.8	23.1	13.6
	s	MPa	131.9	152.3	98.0	13.0	62.0
	cv	%	2.45	5.06	5.5	1.0	8.4

Table A.11 Overview to characterization results of shear strength

	Sign	Unit	RT	150 °C	200 °C	225 °C	250 °C
Strength $R_{12,0.2}$	n	–	4	3	3	3	3
	μ	MPa	60.1	32.2	18.5*	11.4*	7.9*
	$\tilde{\mu}_{RT}$	%	100.0	53.6	30.7	19.0	13.1
	s	MPa	0.59	2.03	1.71*	0.74*	0.81*
	cv	%	0.99	6.31	9.25	6.47	10.28
Strength $R_{12,max}$	μ	MPa	71.3	35.6	18.4*	11.9*	8.5*
	$\tilde{\mu}_{RT}$	%	100.0	49.9	25.8	16.7	11.9
	s	MPa	0.88	0.83	1.46*	0.25*	0.23*
	cv	%	1.23	2.33	7.93	2.10	2.71
Strength $R_{12,ult}$	μ	MPa	113.7	73.7	54.4	36.8	20.9
	$\tilde{\mu}_{RT}$	%	100.0	64.8	47.8	32.4	18.4
	s	MPa	1.55	5.69	1.97	0.42	0.48
	cv	%	1.36	7.71	3.61	1.15	2.26

*Evaluation performed on best-fit polynomials of degree 2

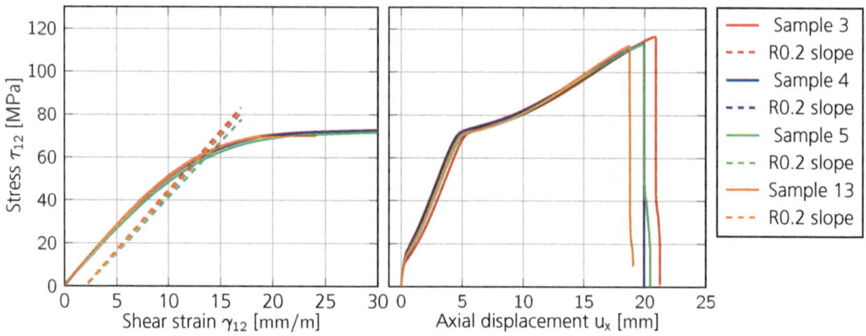

Fig. A.34 In-plane shear characterization at 23 °C

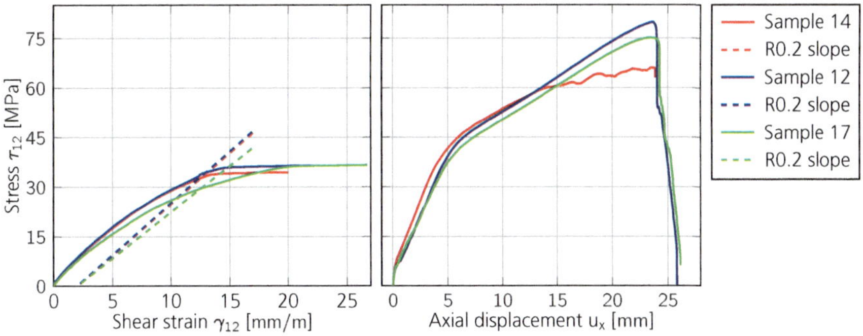

Fig. A.35 In-plane shear characterization at 150 °C

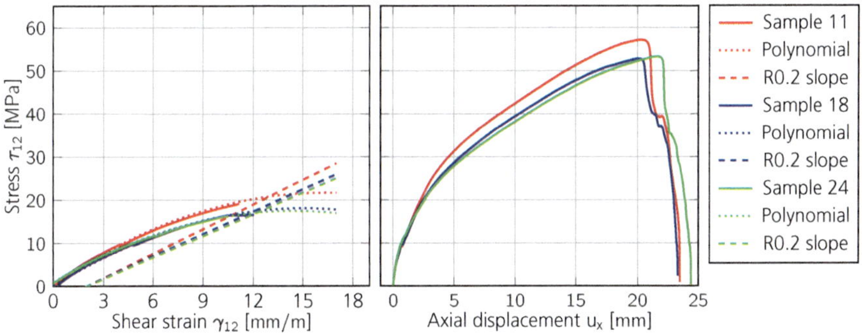

Fig. A.36 In-plane shear characterization at 200 °C

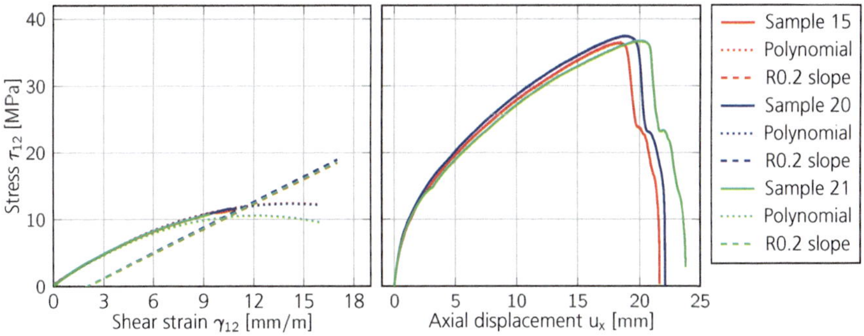

Fig. A.37 In-plane shear characterization at 225 °C

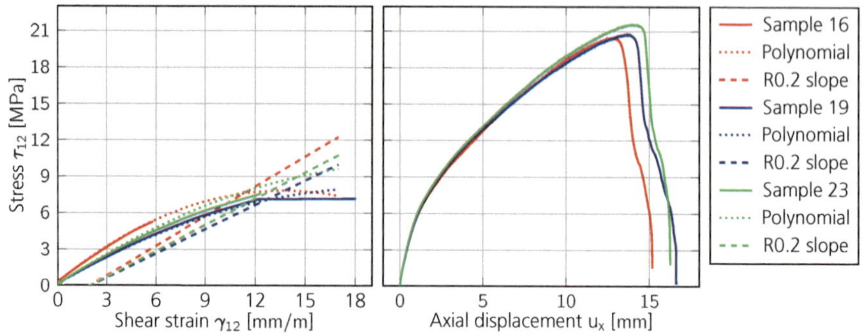

Fig. A.38 In-plane shear characterization at 250 °C

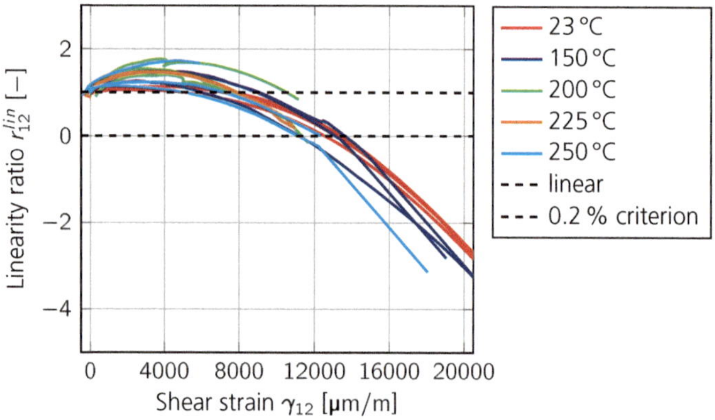

Fig. A.39 Linearity evaluation of the elastic behavior under in-plane shear loading

A.5 Interlaminar Shear Testing

Table A.12 presents an overview to the result of ILSS testing. Figures A.40–A.50 show the measurement curves these results are derived from.

Table A.12 Interlaminar shear testing results

Sign	Unit	23 °C	100 °C	125 °C	150 °C	175 °C	200 °C	225 °C	250 °C	275 °C	300 °C
n	–	7	3	3	3	3	3	3	3	3	4
μ	MPa	75.4	66.2	62.4	60.8	54.5	47.5	31.9	24.2	11.4	9.4
$\tilde{\mu}_{RT}$	%	100	87.9	82.8	80.7	72.3	63.0	42.3	32.2	15.1	12.5
s	MPa	4.2	2.2	4.6	0.7	1.8	1.3	0.3	0.6	0.9	3.7
cv	%	5.6	3.3	7.3	1.1	3.3	2.8	0.9	2.6	8.2	39.2

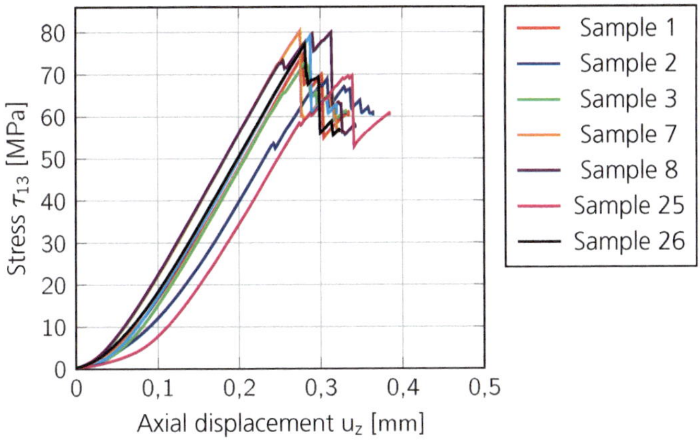

Fig. A.40 Interlaminar shear characterization at 23 °C

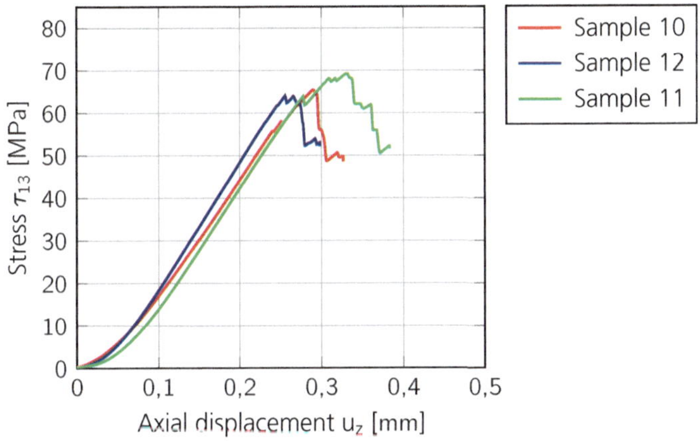

Fig. A.41 Interlaminar shear characterization at 100 °C

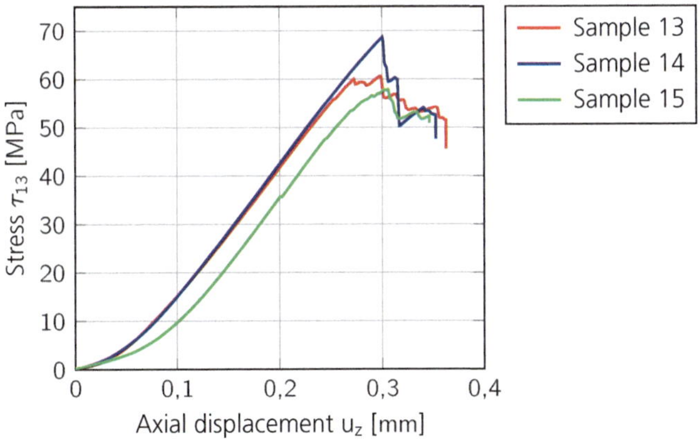

Fig. A.42 Interlaminar shear characterization at 125 °C

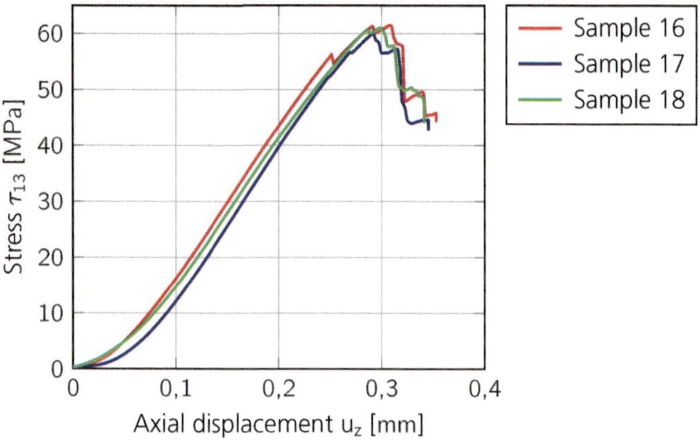

Fig. A.43 Interlaminar shear characterization at 150 °C

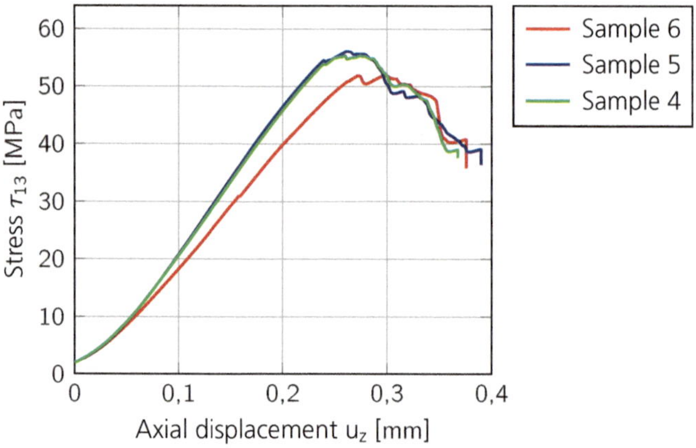

Fig. A.44 Interlaminar shear characterization at 175 °C

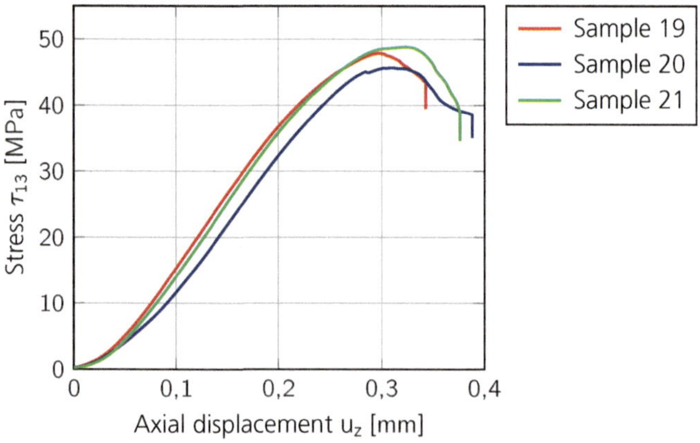

Fig. A.45 Interlaminar shear characterization at 200 °C

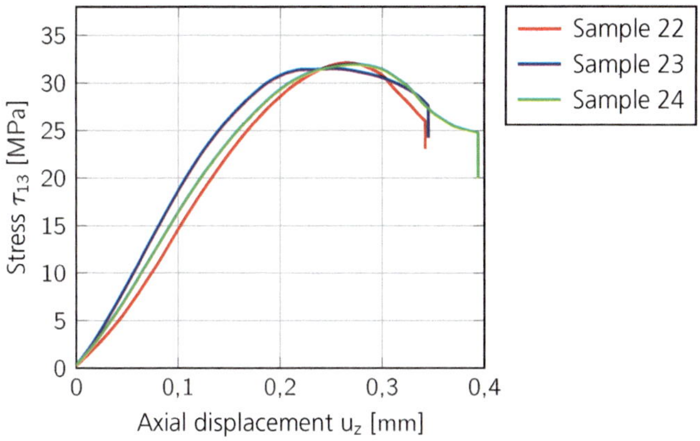

Fig. A.46 Interlaminar shear characterization at 225 °C

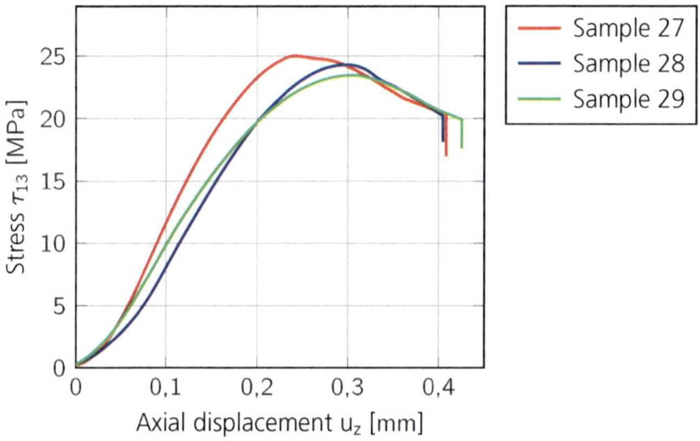

Fig. A.47 Interlaminar shear characterization at 250 °C

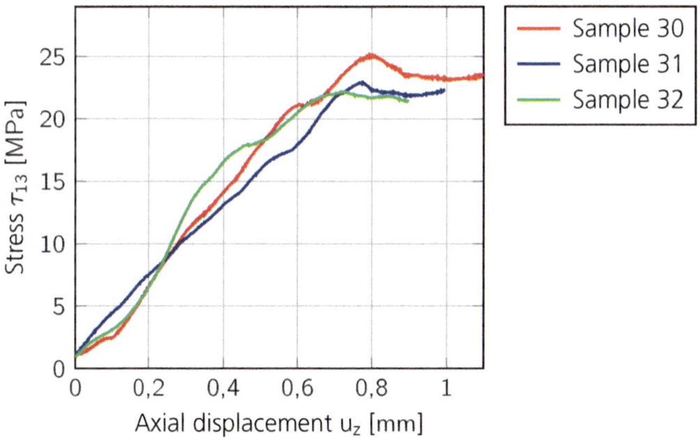

Fig. A.48 Interlaminar shear characterization at 275 °C

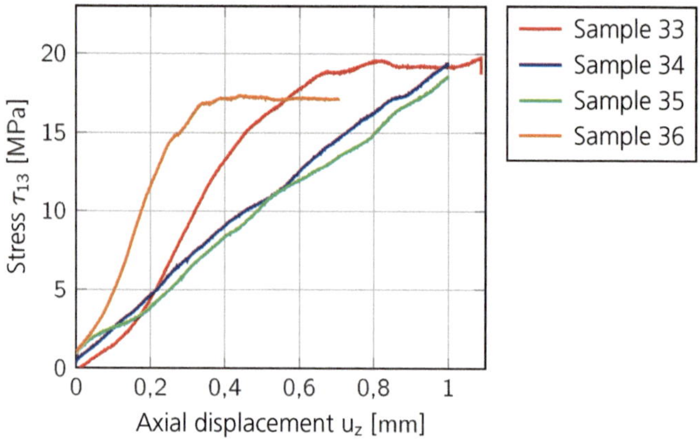

Fig. A.49 Interlaminar shear characterization at 300 °C

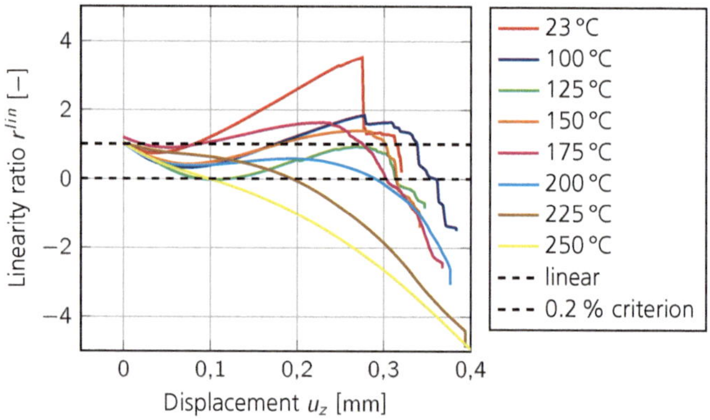

Fig. A.50 Linearity evaluation for ILSS testing. Since ILSS tests did not include strain measurement, the normal deformation is used, as it represents a quantity that correlates to axial strain development. Only a single sample is shown at each temperature

References

1. ASTM International, Standard Test Method for Tensile Properties of Polymer Matrix Composite Materials (2007)
2. DIN-Normenausschuss Luft- und Raumfahrt, Unidirektionale Laminate aus Kohlenstoffasern und Reaktionsharz - Druckversuche parallel zur Faserrichtung (2018)
3. ASTM International, Standard Test Method for In-Plane Shear Response of PMC-Materials by Tensile Test of $a + -45°$-Laminate (2001)
4. Normenstelle Luftfahrt im DIN Deutsches Institut für Normung e.V. DIN EN 2563: Kohlenstofffaserverstärkte Kunststoffe - Unidirektionale Laminate - Bestimmung der scheinbaren interlaminaren Scherfestigkeit (1997)

Appendix B
Failure Criteria

This chapter presents the mathematical relations used to investigate UD FRP lamina failure prediction theories. Depending on theory and approach, there might be different formulations available within the literature. Each theory is presented by its constitutive equations and three plots highlighting the individual failure surfaces on the stress space axis planes and determined using the material model properties determined in Chap. 4. For further information, the author refers to the cited literature.

B.1 Maximum Stress Criterion

According to [1] the structure is undamaged if the following conditions are valid. Thereby, the failure surface is created through equi Invalidity an failure occurs unless at least one condition is equality of stress component and related uniaxial strength component. Invalidity of any condition therefore represents a failed material. The missing interaction of stress components is clearly visible by the rectangular shapes of the failure envelope cuts presented in Fig. B.1.

$$-R_{11c} < \sigma_{11} < R_{11t} \tag{B.1}$$

$$-R_{22c} < \sigma_{22} < R_{22t} \tag{B.2}$$

$$|\tau_{12}| < R_{12} \tag{B.3}$$

M. Liebisch, *Design Load Allowables for Composite Plates Exposed to Thermomechanical Loads*, Mechanics and Adaptronics,
https://doi.org/10.1007/978-3-031-71352-1

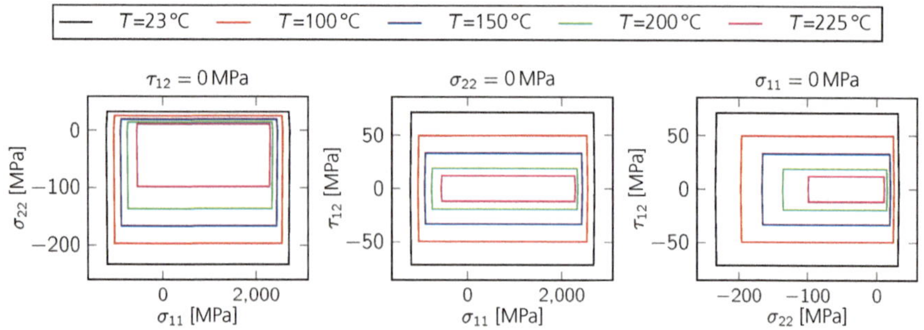

Fig. B.1 Maximum Stress criterion failure surfaces obtained through the stress space axis planes at various temperatures

B.2 Tsai-Wu's Criterion

The formulation shown in Eq. B.4 is based on Mittelstedt [1] and was also referred to by Hashin [2]. Each term shown in Eq. B.4 is generated from the basis formulation shown in Eq. 5.16. The expression represents a reduced form that is applicable for plane-stress conditions and derived for orthotropic material, by introducing the uniaxial strength conditions into the F-terms. By assuming shear stress and normal stress components to be uncoupled, related F-terms vanish and reduce the number of terms [3] simplifying the equation. The interaction term F_{12}^* has to be in the range $[-1; 1]$ and is used with a value of $F_{12}^* = -0.5$ according to [4] (Fig. B.2).

$$\frac{\sigma_{11}^2}{R_{11t}R_{11c}} + \frac{2 \cdot F_{12}^*\sigma_{11}\sigma_{22}}{\sqrt{R_{11t}R_{11c}R_{22t}R_{22c}}} + \frac{\sigma_{22}^2}{R_{22t}R_{22c}} + \frac{\tau_{12}^2}{R_{12}^2} + \sigma_{11}\left(\frac{1}{R_{11t}} - \frac{1}{R_{11c}}\right) + \sigma_{22}\left(\frac{1}{R_{22t}} - \frac{1}{R_{22c}}\right) = 1 \quad \text{(B.4)}$$

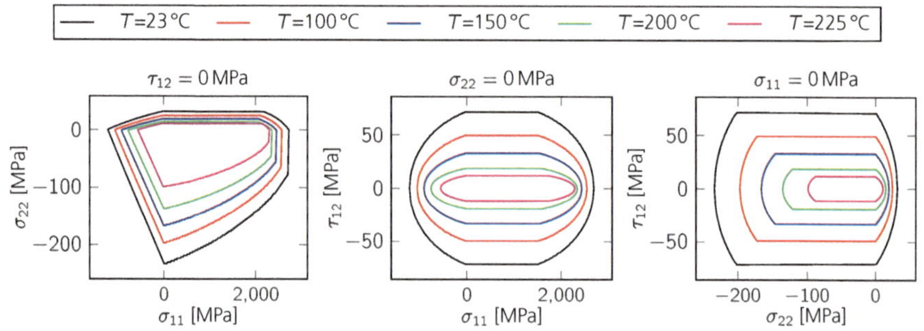

Fig. B.2 Tsai-Wu's criterion failure surfaces obtained through the stress space axis planes at various temperatures

B.3 Hashin's Criterion

The equations shown are taken from [1]. A detailed description of approach and theoretical considerations is provided in Hashins publication [2]. In summary, each equation describes the failure envelope within two quadrants of the stress space (Fig. B.3).

$$\left(\frac{\sigma_{11}}{R_{11t}}\right)^2 + \left(\frac{\tau_{12}}{R_{12}}\right)^2 = 1 \quad \text{for } \sigma_{11} \geq 0 \tag{B.5}$$

$$\frac{-\sigma_{11}}{R_{11c}} = 1 \quad \text{for } \sigma_{11} < 0 \tag{B.6}$$

$$\left(\frac{\sigma_{22}}{R_{22t}}\right)^2 + \left(\frac{\tau_{12}}{R_{12}}\right)^2 = 1 \quad \text{for } \sigma_{22} \geq 0 \tag{B.7}$$

$$\left(\frac{\sigma_{22}}{2R_{12}}\right)^2 + \left(\frac{\tau_{12}}{R_{12}}\right)^2 + \frac{\sigma_{22}}{R_{22c}}\left[\left(\frac{R_{22c}}{2R_{12}}\right)^2 - 1\right] = 1 \quad \text{for } \sigma_{22} < 0 \tag{B.8}$$

B.4 Puck's Criterion

For fiber failure, Puck does not consider severe interaction from transverse stress components or shear stress components. A strain based formulation is applied to consider sensitivity resulting to lateral contraction [5]. For simplicity, the present work refers to a publication of Puck [6] that demonstrates a low variation compared to the *Maximum Stress Criterion* failure conditions at load conditions in fiber direction. Hence, the FF conditions, Eqs. B.9 and B.10, are defined here using stresses instead of strains. The compressive FF considers an empirical shear term, conf. [5] for further

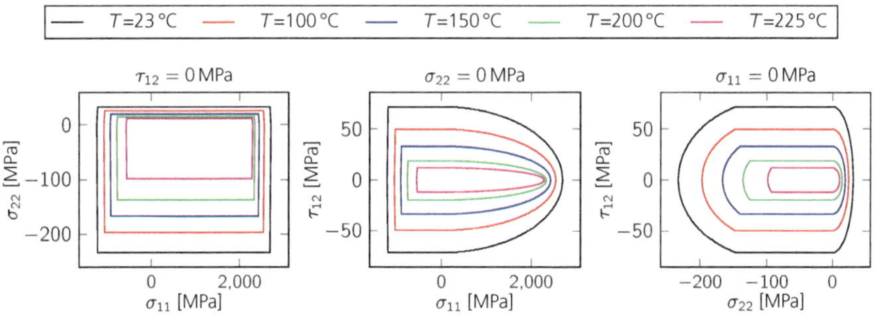

Fig. B.3 Failure surfaces generated using Hashin's criterion and visualized for the stress space axis planes at various temperatures

explanation. The equations equations related to IFF are taken from [5]. In total, five different failure modes are distinguished and addressed by the following equations.

Tensile FF:

$$\sigma_{11} \geq 0 \qquad \frac{\sigma_{11}}{R_{11t}} = 1 \tag{B.9}$$

Compressive FF:

$$\sigma_{11} < 0 \qquad \frac{|\sigma_{11}|}{R_{11t}} + \left(10 \cdot \frac{\tau}{G}\right)^2 = 1 \tag{B.10}$$

IFF Mode A:

$$\theta_{FP} = 0^\circ \tag{B.11}$$

$$\sigma_{22} \geq 0 \quad \sqrt{\left(\frac{\tau_{12}}{R_{12}}\right)^2 + \left(1 - p^{+}_{\perp\parallel}\frac{R_{22t}}{R_{12}}\right)^2 \left(\frac{\sigma_{22}}{R_{22t}}\right)^2} + p^{(+)}_{\perp\parallel}\frac{\sigma_{22}}{R_{12}} = 1 + f_w \tag{B.12}$$

IFF Mode B:

$$\theta_{FP} = 0^\circ \tag{B.13}$$

$$\sigma_{22} < 0$$

$$0 \leq \frac{\sigma_{22}}{\tau_{12}} \leq \frac{R^A_{\perp\perp}}{|\tau_{12c}|} \qquad \frac{1}{R_{12}}\left(\sqrt{\tau_{12}^2 + \left(p^{(-)}_{\perp\parallel}\sigma_{22}\right)^2} + p^{(-)}_{\perp\parallel}\sigma_{22}\right) = 1 + f_w \tag{B.14}$$

IFF Mode C:

$$cos\theta_{FP} = \sqrt{\frac{f_w R^A_{\perp\perp}}{-\sigma_{22}}} \tag{B.15}$$

$$\sigma_{22} < 0$$

$$0 \leq \frac{\tau_{12}}{\sigma_{22}} \leq \frac{|\tau_{12c}|}{R^A_{\perp\perp}} \qquad \frac{R_{22c}}{4\left(R_{12}+p^{(-)}_{\perp\parallel}R^A_{\perp\perp}\right)^2} \cdot \frac{\tau_{12}^2}{-\sigma_{22}} + \frac{(-\sigma_{22})}{R_{22c}} = 1 + f_w \tag{B.16}$$

with

$$R^A_{\perp\perp} = \frac{R_{22c}}{2\left(1+p^{(-)}_{\perp\perp}\right)} \tag{B.17}$$

$$p^{(+)}_{\perp\parallel} = -\left(\frac{d\tau_{12}}{d\sigma_{22}}\right)_{\sigma_{22}=0} \quad \text{for} \quad \sigma_{22} \geq 0 \tag{B.18}$$

$$p^{(-)}_{\perp\parallel} = -\left(\frac{d\tau_{12}}{d\sigma_{22}}\right)_{\sigma_{22}=0} \quad \text{for} \quad \sigma_{22} \leq 0 \tag{B.19}$$

The inclination parameters $p^{(+)}_{\perp\parallel}$ and $p^{(-)}_{\perp\parallel}$ have to be determined by fitting of experimental data [7]. The weakening factor f_w describing the sensitivity of longitudinal stresses on IFF modes is used in similarity to [5]. However, different versions of those terms exist within publications of Puck and a careful study is recommended.

$$f_w = \left(\frac{\sigma_{11}}{1.1 \cdot R_{11}}\right)^n \tag{B.20}$$

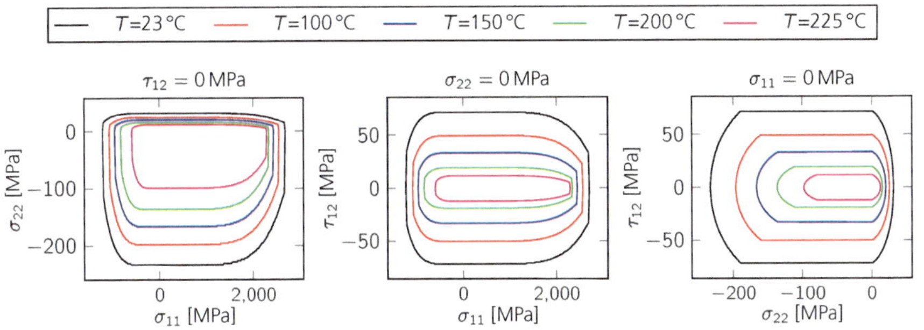

Fig. B.4 Puck's criterion failure surfaces obtained through the stress space axis planes at various temperatures

The exponent n used to determine the weakening factor f_w is comparably sensitive. It is here assumed by $n = 2$ resulting in the failure envelope shape shown in Fig. B.4.

B.5 Cuntze's Criterion

Cuntze [8–11] differentiates five general failure mode's that are likely to occur. For each failure mode, an individual functional relation is derived using invariants. Two failure mode terms represent the typical fiber failure FF approach, i.e. as for the *Maximum Stress Criterion*. Three other terms are related to matrix or interface dominated fracture mode's (IFF). The so called "Rounding-Off Approach" combines all five terms and thereby leads to a continuous failure body. Therein, the exponential coefficient m (usually m = 2.5...3 [9]) is a fitting parameter applied to consider stress interactions in space between the individual failure type domains. μ_{12} represents the inner friction coefficient as explained in detail in [11] with $\mu_{12} = 0.17$ and being applicable in a range of $0.1 < \mu_{12} < 0.3$ [1]. In the present work, a value of $\mu_{12} = 0.2$ is used. The joint equation with the individual failure mode terms is used as following (Fig. B.5):

$$\underbrace{\left(\frac{\sigma_{11} + |\sigma_{11}|}{2R_{11t}}\right)^m}_{FF1} + \underbrace{\left(\frac{-\sigma_{11} + |\sigma_{11}|}{2R_{11c}}\right)^m}_{FF2} + \underbrace{\left(\frac{\sigma_{22} + |\sigma_{22}|}{2R_{22t}}\right)^m}_{IFF1} + \underbrace{\left(\frac{-\sigma_{22} + |\sigma_{22}|}{2R_{22c}}\right)^m}_{IFF2} + \underbrace{\left(\frac{|\tau_{12}|}{R_{12} - \mu_{12}\sigma_{22}}\right)^m}_{IFF3} = 1 \tag{B.21}$$

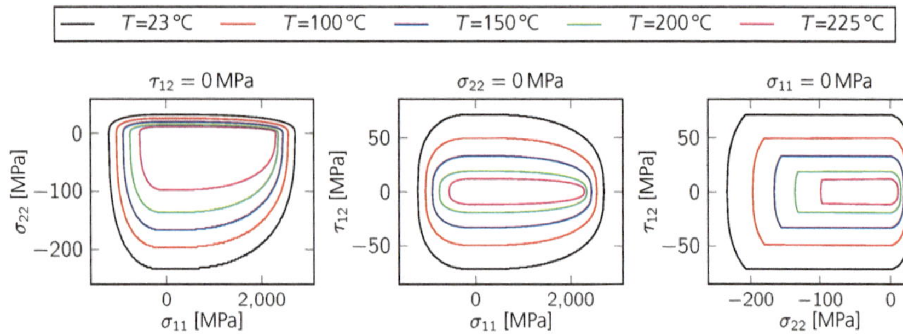

Fig. B.5 Cuntze's criterion failure surfaces obtained through the stress space axis planes at various temperatures

References

1. C. Mittelstedt, W. Becker, Strukturmechanik ebener Laminate, 1st Auflage. (Darmstadt, 2016). ISBN: 978-3-935868-99-0
2. Z. Hashin, Failure criteria for unidirectional fiber composites. J. Appl. Mech. **47**, 329–334 (1980). ISSN: 00218936. https://doi.org/10.1115/1.3153664
3. S.W. Tsai, E.M. Wu, A general theory of strength for anisotropic materials. J. Compos. Mater. **5**(1), 58–80 (1971). ISSN: 0021-9983. https://doi.org/10.1177/002199837100500106
4. S.W. Tsai, J.D.D. Melo, An invariant-based theory of composites. Compos. Sci. Technol. **100**, 237–243 (2014). ISSN: 02663538. https://doi.org/10.1016/j.compscitech.2014.06.017
5. A. Puck, Failure analysis of FRP laminates by means of physically based phe-nomenological models. Compos. Sci. Technol. **58**(7), 1045–1067 (1998). ISSN: 02663538. https://doi.org/10.1016/S0266-3538(96)00140-6
6. A. Puck, H. Schürmann, Failure analysis of FRP laminates by means of physically based phenomenological models. Compos. Sci. Technol. **62**(12–13), 1633–1662 (2002). ISSN: 02663538. https://doi.org/10.1016/S0266-3538(01)00208-1
7. A. Puck, J. Kopp, M. Knops, Guidelines for the determination of the parameters in Puck's action plane strength criterion. Compos. Sci. Technol. **62**(3), 371–378 (2002). ISSN: 02663538. https://doi.org/10.1016/S0266-3538(01)00202-0
8. R.G. Cuntze, The predictive capability of failure mode concept-based strength criteria for multi-directional laminates-part B. Compos. Sci. Technol. **64**(3–4), 487–516 (2004). ISSN: 02663538. https://doi.org/10.1016/S0266-3538(03)00225-2
9. R.G. Cuntze, A. Freund, The predictive capability of failure mode concept-based strength criteria for multidirectional laminates. Compos. Sci. Technol. **64**(3–4), 343–377 (2004). ISSN: 02663538. https://doi.org/10.1016/S0266-3538(03)00218-5
10. R.G. Cuntze, Efficient 3D and 2D failure conditions for UD laminae and their application within the verification of the laminate design. Compos. Sci. Technol. **66**(7–8), 1081–1096 (2006). ISSN: 02663538. https://doi.org/10.1016/j.compscitech.2004.12.046
11. R. Cuntze, The predictive capability of failure mode concept-based strength conditions for laminates composed of unidirectional laminae under static triaxial stress states. J. Compos. Mater. **46**(19–20), 2563–2594 (2012). ISSN: 0021-9983. https://doi.org/10.1177/0021998312449894

Appendix C
Analysis Settings and Results

C.1 Analysis Settings

The deterministic analyses performed according to the description in Sect. 5.1. In the scope pf this thesis, the following details are applied in FEA:

- For the investigations performed in Chap. 6, the thermal analysis is substituted by direct assignment of temperatures as explained in Sect. 5.1.1
  - constant temperature gradient over thickness over the full structure
  - linear temperature gradient over thickness
- Boundary conditions of mechanical analysis, conf. Fig. 5.4
  - edge1, edge3: $u = z = u_r = z_r = 0$
  - edge2, edge4: $y = z = y_r = z_r = 0$
  - Axial loading is applied via shifting edge1: $u_{ax} > 0$
- Geometrical imperfection is considered by the third eigenmode and an amplitude of 0.001
- Temperature and section points are selected to be coincident. The temperature points are obtained by equidistant spacing over the full laminate and the section points are defined using Simpson's rule (3 per layer, one at each layer surface and one at its center) this leads to:
  - The CP laminate used 9 temperature points and 12 sections points
  - The MAP laminate used 17 temperature points and 24 section points

The evaluation is performed in a postprocessing step after finishing FEA. For failure evaluation, FPF is employed and the results in Chap. 6 use Puck's criteria for IFF and FF, as described in Sect. 5.3. Progressive damage or interaction between IFF and FF are not considered.

M. Liebisch, *Design Load Allowables for Composite Plates Exposed to Thermomechanical Loads*, Mechanics and Adaptronics,
https://doi.org/10.1007/978-3-031-71352-1

For the probabilistic analysis, the thermal parameter space is reduced to $X^{th} = \{T^{max}, T^{avg}\}$. The distribution of training points is irregular, since these parameters are based on the parameters of Eq. 5.2 and the thickness-wise temperature gradient applied. Surrogate modelling and probabilistic analysis is performed based on these thermal parameters and the resulting design parameters

- Each Monte Carlo Analysis used 10000 points
- PDFs of the thermal input parameters Xth are defined as following
  - T^{avg} is assumed by normal distribution
  - T^{max} is assumed by a general extreme value distribution
  - The PDFs of T^{max} and T^{avg} do not correlate.

C.2 Cross-Ply Laminate: Structural Behavior

Figure C.1 presents the variation of the stability behavior in dependence of temperature. Figures C.2 and C.3 highlight the behavior of the structure on exemplaric temperature distributions.

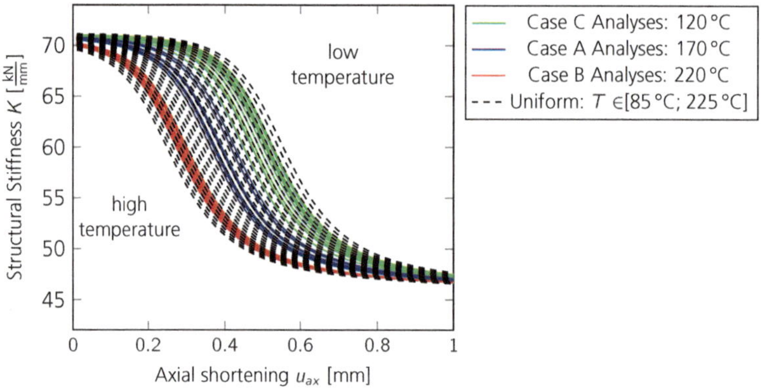

Fig. C.1 Structural stiffness of the CP Laminate for different thermal states. The distributed analyses (colored lines) are represented by 10 exemplaric curves for each thermal case. The uniform analyses (black dashed lines) are represented in steps of 5 K

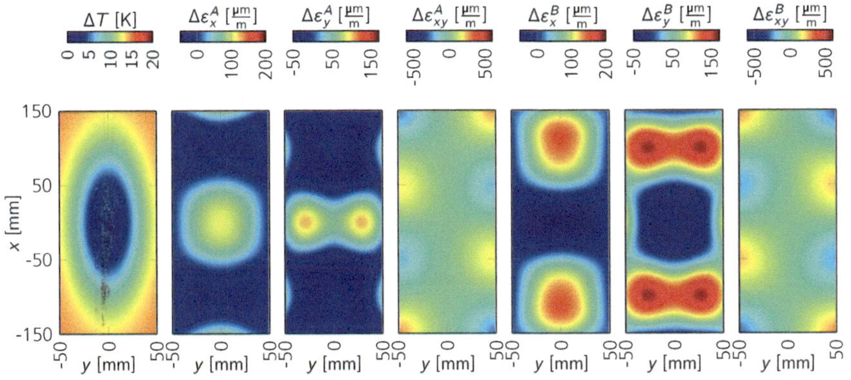

Fig. C.2 Difference of the global strain response between uniform reference result (170 °C) and an exemplaric DiA result for the CP laminate: Case A ($T^{avg} = 157.6\,°C$, $T^{max} = 170.0\,°C$) at 40.1 kN axial loading. Surfaces are indicated by superscripts A (heated surface) and B (opposite side). The left chart shows the temperature difference at the heated surface. The other plots refer to the strains differences at both surfaces are generated by subtracting the distributed analysis quantity from the uniform analysis quantity

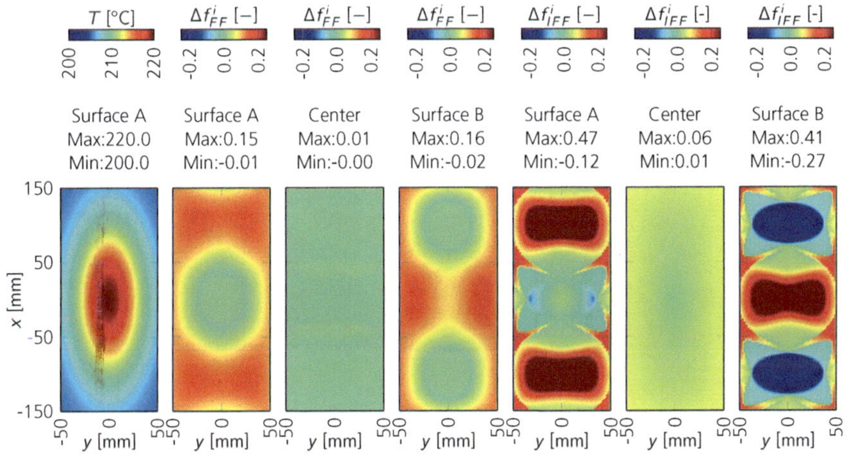

Fig. C.3 Difference of the resulting failure index distribution between uniform reference (220 °C) result and distributed analysis result for the CP laminate, Case B ($T^{avg} = 207.6\,°C$, $T^{max} = 220.0\,°C$) at 45.8 kN axial loading. The plots are generated by subtracting the distributed analysis quantity from the uniform analysis quantity

C.3 Cross-Ply Laminate: Surrogate Response

The Figs. C.4 to C.6 present the surrogate response over the parameter space for the investigated design criterions.

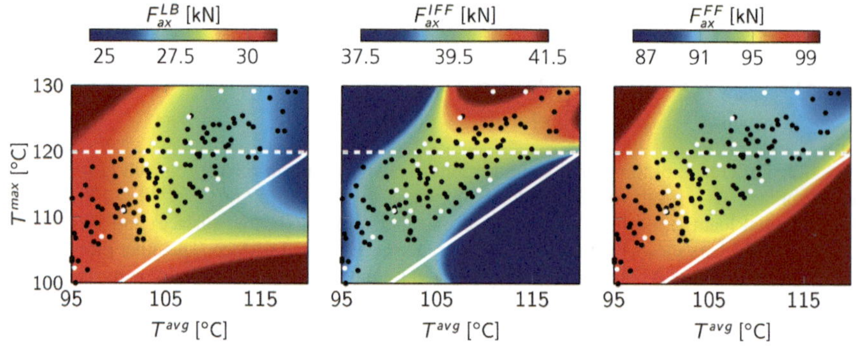

Fig. C.4 Surrogate responses for the CP Laminate Case C (120 °C)

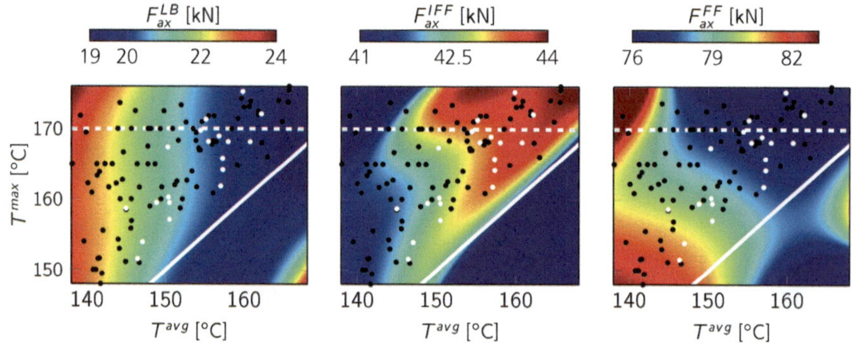

Fig. C.5 Surrogate responses for the CP Laminate Case A (170 °C)

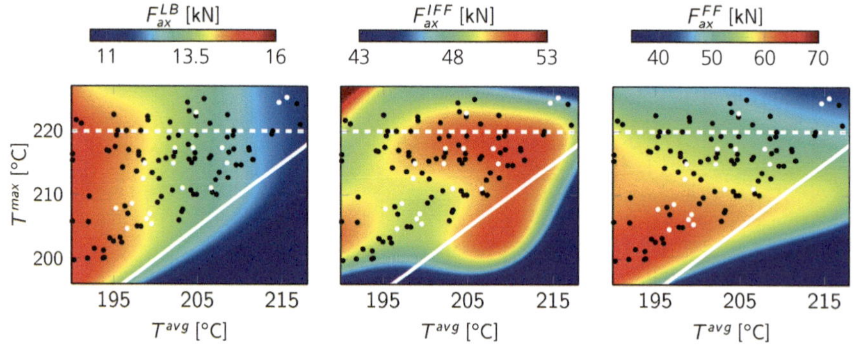

Fig. C.6 Surrogate responses for the CP Laminate Case B (220 °C)

C.4 MAP Laminate: Structural Stiffness

Figure C.7 presents the variation of the stability behavior in dependence of temperature. Figure C.8 highlights the temperature distributions leading to varying stability behavior and discussed in Fig. 6.8.

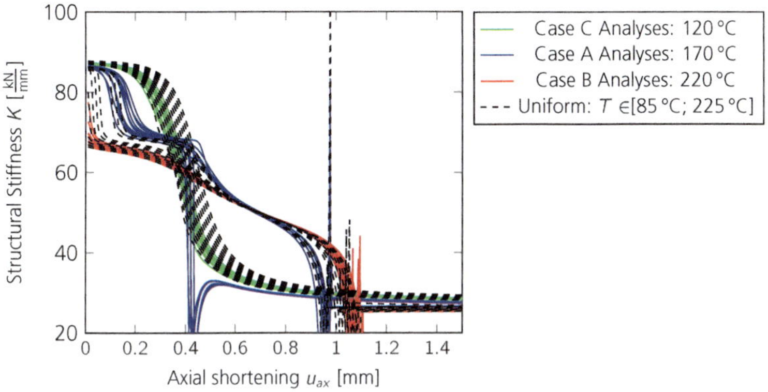

Fig. C.7 Structural stiffness of the MAP laminate for different thermal states. The distributed analyses (colored lines) are represented by 10 exemplaric curves for each thermal case. The uniform analyses (black dashed lines) are represented in steps of 5 K

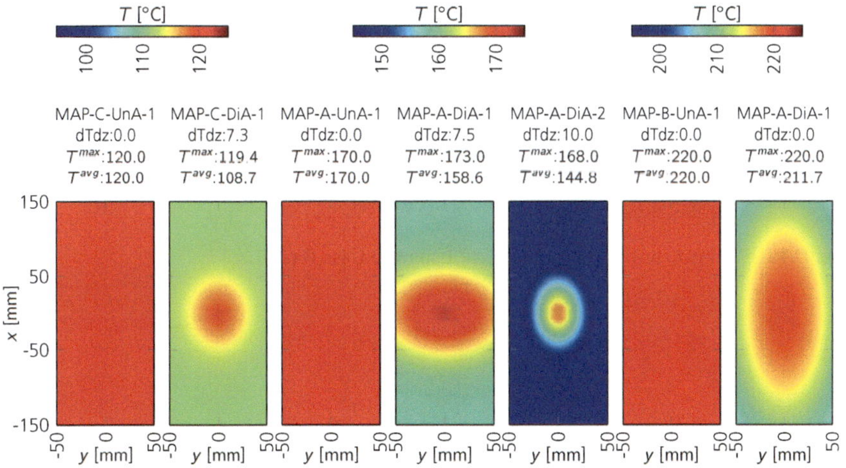

Fig. C.8 Temperature distributions considered in specific analyses of the MAP laminate. They refer to the study explained in Fig. 6.8

C.5 MAP Laminate: Surrogate Response

The Figs. C.9 and C.10 present the surrogate response over the parameter space for the investigated design criterions.

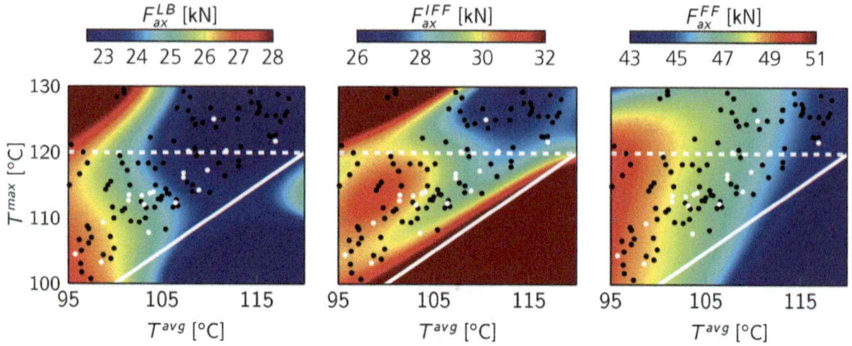

Fig. C.9 Surrogate responses for the MAP Laminate Case C (120 °C)

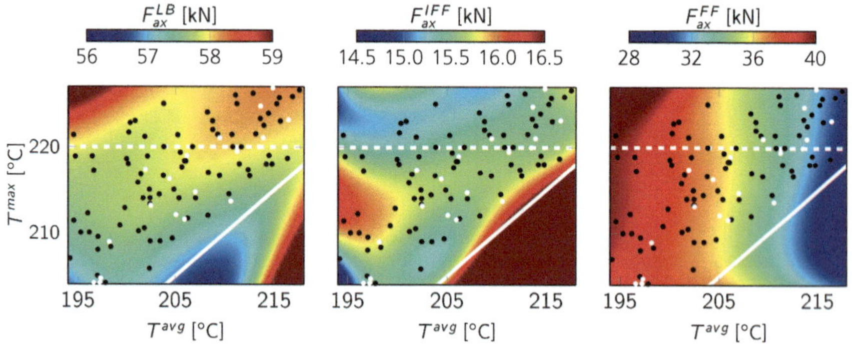

Fig. C.10 Surrogate responses for the MAP Laminate Case B (220 °C)

GPSR Compliance

The European Union's (EU) General Product Safety Regulation (GPSR) is a set of rules that requires consumer products to be safe and our obligations to ensure this.

If you have any concerns about our products, you can contact us on ProductSafety@springernature.com

In case Publisher is established outside the EU, the EU authorized representative is:

Springer Nature Customer Service Center GmbH
Europaplatz 3
69115 Heidelberg, Germany

Batch number: 10726060

Printed by Printforce, the Netherlands